Environmental Disaster Management

Environmental Disaster Management

Vierah Hulley

Delve
PUBLISHING
www.delvepublishing.com

Environmental Disaster Management

Vierah Hulley

Delve Publishing
224 Shoreacres Road
Burlington, ON L7L 2H2
Canada
www.delvepublishing.com
Email: orders@arclereducation.com

ISBN: 978-1-77407-753-5 (Hardcover)

ABOUT THE AUTHOR

Vierah Hulley obtained her Ph.D. degree in Geohydrology from the University of the Free State, South Africa, in 2013. Her expertise is in Environmental and Earth Sciences and she has an extensive vocational background in the areas of natural resources management, environmental geochemistry, contaminated site remediation, sustainable development, and spatial analytics. She is the founder and Managing Director of HL Nexus: The Data Imaginarium; a company dedicated to optimizing natural resources management with data science.

TABLE OF CONTENTS

LIST OF FIGURES

LIST OF ABBREVIATIONS

APA	American Planning Association
APELL	awareness and preparedness for emergencies at local level
ASAL	arid and semiarid land
BNPB	National Board for Disaster Management
BSL	biosafety level
CCA	common country assessment
CDC	Center for Disease Control
COG	Continuity of Government
COOP	continuity of operations plans
DRM	disaster risk management
DRR	disaster risk reduction
EIA	environmental impact assessment
EMS	emergency medical services
EMUs	emergency medical units
FEAT	flash environmental assessment tool
FEMA	Federal Emergency Management Agency
GIS/RS	geographic information systems/remote sensing
IPCC	Intergovernmental Panel on Climate Change
JDLNA	joint damages losses and needs assessment
MSHA	Mine Safety Health Administration
NIST	National Institute of Standards and Technology
NOAA	National Oceanic Atmospheric Administration
OCHA	office for coordination of humanitarian affairs
PAHs	polycyclic aromatic hydrocarbons
PCEA	post-conflict environmental assessment
PCNAs	post-conflict needs assessments
PDNAs	post-disaster needs assessments
RRTs	rapid response teams

RSFs	recovering support functions or
SA	strategic assessment
SEA	strategic environmental assessment
SoE	state of environment reporting
SOE	state-owned enterprise or
UNESCO	UN's Educational, Scientific, and Cultural Organization
UV	ultraviolet
WPUs	water purification units

PREFACE

An environmental disaster can be described as a disastrous event caused primarily by human activity, resulting in harmful effects on the biosphere. In most situations, environmental disasters come about due to human error, lack of foresight, accident, greed, sheer incompetence, and poor industrial practices. Meaning, without some form of human involvement they wouldn't have occurred. In some cases, the culprits deny involvement, and may also escape any effects.

In the agricultural industry, a classic case of a disaster is the increased salinity of soils within hot climates. Considering the need for producing more food, warm climates seem perfect for growing European-style crops, once the existing foliage has been removed. One provision is that there should be sufficient water, and for this irrigation, programs and deep wells normally are the solution. However, as has been established in some regions, if this process isn't properly managed the resulting salinity may cause the land to become sterile and practically useless.

Industrial pollution is another cause of environmental disasters in many countries. For instance, during the mid-80s a methyl isocyanate leakage that has come to be known as the Bhopal disaster-affected 1000s of people, and led to 22,000 deaths including various genetic anomalies that will persist for generations. The main causes of this tragedy were found to be slackness, corruption, and total disregard for safety measures.

Not long ago, a drilling rig along the Gulf of Mexico suddenly exploded after the safety valve developed to prevent oil spillage remarkably failed. It took several months for the leakage to be re-sealed, during which time tons of oil had already poured into the surrounding sea. The ensuing pollution wasn't just from the petroleum, but from chemical compounds which were applied to dissipate it. Entire ecosystems were damaged together with the lives of many local people. Besides, many endangered species aren't expected to recuperate at all.

Introducing alien species of plants or animals to a different location can also cause environmental problems. For instance, some years back dozens of imported English bush rabbits were introduced in Australia to help a local farmer go hunting. But this proved disastrous years down the line, since the native Australian rabbit populations grew from the original 12 to now more than 200 million as estimated.

Apart from being responsible for devastating massive acreages of cropland and grazing fields, they have also consumed foliage that was meant for other species thereby leading to their rapid decline. The invasive rabbits also chew on young trees and prevent them from growing, something that has contributed to the barren landscape experienced in Australia today.

This volume discusses various issues of concern in the management of environmental disasters. Chapter 1 is an introduction to what a disaster is, Chapter 2 talks about factors leading to environmental disasters, Chapter 3 consequences of environmental disaster, Chapter 4 factors of vulnerability to environmental disasters, Chapter 5 environmental disaster preparedness, Chapter 6 environmental disaster recovery (rehabilitation and reconstruction), Chapter 7 environmental disaster mitigation, Chapter 8 environmental health management, Chapter 9 worst environmental disasters in developing countries: case studies in Africa, and Chapter 10 worst man-made environmental disasters in American history.

1

What is a Disaster?

CONTENTS

1.1. INTRODUCTION

A disaster can be defined as a severe disruption happening over a brief or extended period of time, causing prevalent human, economic, material, and environmental loss that surpasses the capacity of an affected community, or society to comfortably cope using their own resources (Figure 1.1).

Figure 1.1: Children from a developing country playing next to a collapsed building.

Source: https://www.ifrc.org/en/what-we-do/disaster-management/about-disasters/what-is-a-disaster/.

Underdeveloped countries normally suffer the biggest costs whenever a disaster strikes-well over 95% of all deaths resulting from hazards happen in developing nations, and losses coming from natural disasters are 20 times more profound in developing regions compared to industrialized nations.

Historically, the term disaster is borrowed from the French dialect "désastre" and an Old Italian phrase "disastro," which in turn has been derived from Ancient Greek. Disasters are normally divided into man-made and natural types; however, complex disasters in which there isn't a particular root cause may also happen.

One disaster may cause secondary disasters which escalate the overall impact. A good example of this is an earthquake; it's a natural disaster which consequently causes environmental damage as a result of tsunami and coastal flooding. A number of manufactured disasters furthermore have been tied back to nature. Besides, researchers also distinguish between recurrent events like seasonal flooding and those regarded as unpredictable.

Natural disasters cause a number of deaths and this number varies yearly. There are years that can pass with very few deaths before a disaster comes and claims a lot of lives. According to the United Nations, natural disasters claim 60,000 lives yearly on average and this is representative of 0.1% of deaths globally. This numbers can be low as 10,000 yearly and these accounts for 0.01% of deaths in total at the same time there are disasters that claim over 200,000 lives such as famine and drought, tsunamis, and earthquakes and this in those years account for approximately 0.4% of death. Some of the natural disasters are not preventable and therefore result in large and multiple losses. Their data has also shown significant decline of the death toll from natural disasters from the 1900s and through the 2000s (United Nations, 2010).

Additionally, natural disasters have environmental impact and can trigger significant loss of life, injuries, and other health impacts, property destruction, loss of occupations and services, economic, and social disruption, as well as ecosystem damage. Different phenomena such as earthquakes, volcanic eruptions, landslides, floods, blizzards, hurricanes, tornadoes, cyclones, and tsunamis can all be categorized as natural disasters which kill many people and damage billions of dollars' worth of environment and property every year.

Nevertheless, the rapid progress of the globe's population and its amplified concentration in dangerous environments has intensified both the frequency as well as the severity of disasters, some of which are catastrophic. Developing countries have to deal with tropical climates and uneven landforms, together with deforestation, unforeseen growth proliferation, and other conditions that make them more vulnerable. Typically, developing nations suffer more constantly from environmental disasters because of ineffective communication, together with inadequate budgetary provision for disaster avoidance and management.

Plane crashes and terror attacks are common examples of human-made environmental disasters: causing pollution, killing people and destroying property. Similarly, human-instigated disasters emerge as a result of technological and manmade hazards. Some common examples are; transport accidents, manufacturing accidents, public stampedes, fires, oil spillages, terrorist attacks, and nuclear radiations. Wars and intentional attacks can also be grouped in this category. Some other forms of instigated disasters are of universal scale like nuclear war, bioterrorism, and global warming.

1.2. DEFINITION OF SOME ENVIRONMENTAL DISASTERS

1.2.1. Natural

- **Avalanche:** The abrupt, drastic drift of a mass of snow going down a mountain slope under the force of gravity. Avalanches are triggered due to the following: overloading, temperature, slope angle, snowpack conditions, and vibration. It typically happens when natural triggers, like loading from fresh snow or rain, and artificial triggers, like explosives or ice skiers, overloading of snowpack Shutoff facilities. For survival, evacuation of any surrounding buildings is essential. It's also important to determine potential impacts on equipment, facilities, or any other possible disruption.
- **Earthquake:** The abrupt shaking of earth's top layer, triggered by underground volcanic pressures of breaking and unstable rock below the earth's surface. For safety, quick evacuation of building would be necessary. It's also important to determine the impact on equipment or facilities including any possible disruption.
- **Heat Wave:** An extensive period of extremely hot weather in comparison to the normal weather patterns of a region, plus relative to regular temperatures for the weather period. The best way forward is listening to weather advisories; for example, it's recommended to power-off every possible server following a graceful shutdown in case there's looming probability of power failure. It's also important to shut off the central electric circuit typically found in the first-floor basement.

1.2.2. Human Caused

- **Manmade Bioterrorism:** The deliberate release or dispensation of biological contaminants as a way of coercion. For this, it's essential to get information promptly from public health experts through news media, regarding the appropriate course of action.
- **Civil Unrest:** This is simply a disturbance instigated by a number of people which may be in the form of sit-ins, including other types of obstacles, riots, sabotage, plus other kinds of crimes, whose intention is demonstrating chaos to members of the public

or government, though can escalate into violence when not properly managed.

- **Toxic Material Spillage:** The discharge of liquids, solids, or gases which may harm individuals, living organisms, property, and the environment in general, from a container or bag. For such a scenario, leave the affected region and contact the regional chemical disaster department for assistance.

The regional emergency medical services (EMS) line is the primary contact during a toxic material spillage. Nuclear and radioactive material accidents also fall within this category. It covers any event involving substantial release of radioactive elements to the environment or reactor core melt-down that causes major undesirable effects to people or surrounding environment.

1.3. TYPES OF DISASTERS

What Classifies a Disaster? Only the disasters which can fulfill either of the situations mentioned below are captured: (a) 10 or more individuals reported killed, (b) 100 or more persons reported affected, (c) Declaration of an emergency state, and (d) Call for global assistance (Dufresne, 2019).

Fatalities basically are the figure of killed persons from a given disaster, on the contrary, killed is described as individuals who have been confirmed dead, or individuals missing or presumed dead. Individuals affected, on the other hand, are overall figure of Total affected depending on the established definition criteria (Dufresne, 2019).

1.3.1. Natural Disasters

They are naturally happening physical phenomena that are caused by quick or slow onset actions, which have immediate effects on human health as well as secondary effects that cause additional death and suffering. The disasters may be grouped as follows:

- **Geophysical:** Earthquakes fall under this category. Almost every year, the globe experiences natural disasters caused by this phenomenal. When they occur in highly populated regions, many people may potentially lose their lives. One recent case is in Italy where a 2016 earthquake killed well over 200 people. Earthquakes occur when the earth churns out pent-up energy, causing the upper ground to rumble (Babar, 2007).

Earth's outer layer or crust consists of considerably large parcels of land, known as tectonic plates. Many earthquakes happen when these plates' margins scrub against one another in some form. The main causes of earthquakes are:

- Tectonic plate movements;
- Volcanic earthquakes;
- Geological faults (Intra-plate earthquakes);
- Man-made.

Tectonic plates furthermore develop mountains, especially when they converge against one another. When mountains are pushed up, earthquakes can sometimes be felt. Occasionally, people also cause earthquakes whenever they perform mine blasts and nuclear tests (Dufresne, 2019; United Nations, 2010).

- **Volcanic Earthquakes:** Volcanically induced earthquakes have the prospective of causing cracks, ground distortion, and destruction of manmade structures. They normally are much smaller compared to earthquakes triggered by non-volcanic sources. The two forms of volcanically produced earthquakes are volcano-tectonic and lengthy period earthquakes. Volcano-tectonic earthquakes are, basically, caused by slide on a fault around a volcano. Normally volcanoes occur in zones of crustal weakness, while the load of the volcano itself contributes to the local pressure.
- **Intraplate Earthquakes:** Around 5% of earthquakes occur within a plate, far from plate boundaries. The intraplate earthquakes are due to stresses inside a plate. Because plates move across a spherical surface, areas of weakness may be developed. Intraplate earthquakes occur along these regions of weakness. The earthquakes can occur along ancient cracks or rift zones.

As for hurricanes, cyclones, and typhoons, they refer to a particular weather phenomenon, whereby a considerably massive storm spins in circles destroying anything in its path. From outside, you'll watch the storm cloud spinning in a spiral motion, before touching down on the surface, before reaching up toward the sky. Whenever a storm hits wind speeds of more than 74 miles (119 km) an hour, the winds may become classified as typhoon, hurricane or cyclone based on where a given storm is situated.

The storm is known as a hurricane when it occurs mostly in the northern Pacific or Atlantic. In case the storm happens in places such as the north-western territory of the Pacific, it becomes a typhoon. As for storms in the Indian Ocean and southwestern Pacific, they are known as a cyclone (Dufresne, 2019).

- **Climatological:** These disasters can be defined as events that are caused by long-term/micro to macro scale procedures in the scope from intra-seasonal up to multi-decadal weather variability. The events may further be classified into Drought, Wildfire, and Extreme Temperature. The latter events include cold waves, heat waves and harsh winter conditions including (icing, snow pressure, avalanche, and freezing rain). For wildfire, these are forest fires or land fires including (grass, scrub, and bush) among other similar circumstances.
- **Hydrological:** The most common examples are avalanches and floods. An avalanche occurs when there's a pack of snow, rocks or ice falling swiftly down a slope, typically a mountainside. It's possible that you may have already watched movies whereby somebody is standing near a mountain, having a pack of snow at the top. They yell out something then snow comes tumbling down the hill toward them. This is what's known as an avalanche. An avalanche has the potential of burying humans and animals alive.

Flooding, on the contrary, is a water overflow covering a wide area of land. The overflow may destroy buildings, cars or other items along path. For severe flooding, water may seep into homes and totally cover them, destroying everything. Besides, whenever people become caught up in a flood situation, they may be washed away through the flood before being drowned (Dufresne, 2019).

- **Meteorological Disasters:** Meteorological disasters are typically caused by excess weather like rain, snow, excess heat or cold, drought, ice, or wind among other factors. Violent, abrupt, and damaging adjustments in the environment related to, created from, or affecting earth's atmosphere, especially the weather-forming procedures. Some common examples of weather-based disasters are cyclones, blizzard, droughts, heat waves, hurricanes, hailstorms, floods (triggered by rain) and even tornadoes (Babar, 2007).

Furthermore, extra-tropical cyclones, commonly referred to as mid-range cyclones, refer to a group of cyclones identified as synoptic scope, low weather structures that happen within the core latitudes of a planet (away from the tropics) without having any tropical features, are associated with fronts or horizontal levels in temperature, and temperatures also known "baroclinic zones."

Drought is a harsh environmental condition that leads to lack of water due to extreme heart, caused by significantly lower rainfall than average over the chronic amount. It's characterized by hot and dry winds, elevated temperatures, and ensuing evaporation of wetness. Additionally, hailstorms are raindrops which come as ice, rather than melting before they reach the bottom, which is a rather destructive storm hit (Babar, 2007).

A tornado basically is a violent, risky, and rotating pillar of air which is churned as it moves across the earth. It's furthermore called a twister or cyclone, even though the term cyclone is used in meteorology on a wider sense, referring to talking over with whichever closed low circulation. Besides, tornadoes are usually in various shapes and sizes; nevertheless, they're typically within the sequence of a primary condensation funnel, where it's also encircled with a haze of trash or dirt.

Many tornadoes have elevated wind speeds; however 177 km/h are simply around 250 ft (80 m) across, plus travel several kilometers before dissipating. Additionally, the foremost excess tornadoes will achieve wind speeds of up to 480 km/h, stretching over 3 kilometers across, plus remain at the bottom for several miles (possibly over a 100 km) (Babar, 2007).

- **Biological:** A biohazard is a biological compound that poses a significant risk to the wellbeing of living organisms, especially humans. This may comprise of a micro-organism, virus, or even toxin which may badly affect human health. The biohazard may furthermore be a substance that's damaging to other animals.

The official biohazard symbol was first developed in 1966 courtesy of Charles Baldwin, who is an environmental health engineer employed with Dow Chemical Company. It's used in labeling of biological compounds that carry a substantial health risk, such as viral samples and second-hand hypodermic needles (Babar, 2007).

The U.S. Center for Disease Controlling and Prevention (CDC) classifies different illnesses according to the levels of bio-hazard: Level 1 is minimum risk while Level 4 is extreme hazard. Laboratories and related facilities are generally classified as biosafety level (BSL) 1 to 4 or even P1 through to P4 (Protection or Pathogen Levels).

- **Biohazard Level 1:** Microbes and viruses such as; *Canine hepatitis*, *Varicella (chickenpox)*, *Bacillus subtilis,* and *Escherichia coli*, including some forms of cell culture as well as non-infectious bacteria. For this level, precautions are taken against any biohazardous compounds in question are few, most likely consisting of gloves and other kinds of facial safeguards.
- **Biohazard Level 2:** Microbes and viruses which cause just mild infection to people, or are hard to contract through aerosol in the lab setting, like hepatitis A, B, or C, certain strains of influenza A, Lyme disorder, mumps, measles, salmonella, dengue fever and even HIV. Routine diagnostic task with clinical specimens may be performed safely on BSL 2, through using effective BSL 2 activities and procedures. Additionally, research work (such as co-cultivation, viral replication studies, and manipulations that involve concentrated virus) may be performed in a typical BSL-2 (P2) amenity, using BSL-3 systems and procedures.
- **Biohazard Level 3:** Microbes and viruses which may cause intense to fatal illnesses in humans, though for which vaccines and other alternative treatments exist, like anthrax, Venezuelan equine encephalitis, West Nile virus and SARS virus
- **Biohazard Level 4:** These are viruses responsible for causing severe to fatal illness in people, plus for which vaccines and other treatments aren't available, like Marburg virus, Bolivian hemorrhagic fever, Lassa fever virus and Ebola virus.

Variola virus or (smallpox) is basically an agent that's handled at BSL-4, regardless of the presence of a vaccine, since it's been exterminated. When handling biological hazards within this level, it's important to apply a positive force personnel suit with segregated air-supply which is mandatory.

The entrance and exits of Level 4 bio lab may contain several showers, vacuum rooms, ultraviolet (UV) light-room, mechanized detection system, and different other safety pre-cautions developed to destroy various biohazard traces.

Furthermore, various airlocks may be employed or electronically secured for preventing doors from either opening at equal moments. Similarly, all water and air service entering and coming from the BSL IV (P4) lab shall undergo the same decontamination processes, in order to remove any possibility of accidental discharge. Presently, there are no microbes classified under this level.

1.3.2. Manmade Disasters

Manmade disasters as defined by the Red Cross and Red Crescent humanitarian groups are events which are caused by people and happen in or near human habitations. For the most part, they are caused by environmental and technological emergencies such as accidents and environmental degradation (Paul, 2011) (Figure 1.2).

Figure 1.2: A bomb explosion is an example of a manmade disaster.

Source: https://www.insider.com/worst-modern-manmade-disasters-world-environment-day-2019-5.

These disasters arise from human activities like explosions, release of toxic compounds and radioactive materials, fires, crashes, dam failure and building or bridge collapse among other related problems. Technological disasters usually involve a lot of casualties in comparison to natural disasters, despite having the same level of energy release.

Additionally, both man-made and technological catastrophes are unpredictable, and can spread along vast geographical regions; they could be unpreventable and might have limited physical destruction but long-term impacts. They're also way more difficult for communities to deal with or victims to accept.

For technological calamities, there are matters of blame pertained plus the community spends lots of time debating who was liable and what errors were done. Gradually, agencies involved in calamities and management of these are concerned about the interactions between humans and nature, which may be complicated and further aggravate disasters (Paul, 2011).

Communities where industrial sites are situated or through which toxic material pass through roadway, rail, and pipeline are especially at risk of technological adversities. Injuries may happen to personnel at the site, rapid responders trying to bring the matter under control and offering emergency medical assistance, plus to inhabitants in the community. Individuals with pre-existing health conditions, like lung or heart illnesses, may also suffer increased risk of negative health effects if exposed to hazardous skin conditions, besides lung damage may arise from exposure to particular agents.

Ensuring that local industries apply basic safety processes can considerably decrease negative health effects, coming from accidental discharge of toxins. For emergency preparedness, it includes the capacity of pre-hospital as well as hospital structures to take care of patients who are exposed to industry agents, including training of medical staffs to work in polluted environments, plus the stockpiling of private protective gear for responders-it's essential for offering care after industrial accidents (Paul, 2011).

1.3.2.1. Environmental Disasters Caused Through Anthropogenic Influence

Earth's makeup comprises of a lithosphere, atmosphere, biosphere, and hydrosphere. Human actions affect most of the processes which shape these systems. Of specific concern to disaster risk communities are the anthropogenic impacts on the occurrence, rate, and impact of natural hazards, like earthquakes, floods, landslides, sinkholes, and subsidence.

Anthropogenic activities are described as being deliberate and often non-malicious human activities that negatively affect the environment. Some common examples are groundwater abstraction, vegetation removal, subsurface mining, infrastructure (loading), and chemical explosions.

1.3.3. Other Classifications of Disasters

- **Complex Emergencies:** Some calamities may arise from several hazards, or, even, from a complex blend of natural and manmade causes, involving an interruption of authority, looting, as well as attacks targeting strategic installations that disrupts the livelihoods of people, such as during conflict conditions and war. Complex emergencies may cover matters such as, food insecurity, armed conflicts, displaced populations, and epidemics. These emergencies are normally characterized by issues like extensive violence, displacements of people, loss of life, extensive damage to societies as well as economies (Paul, 2011).
- **Pandemic Emergencies:** Pandemic is basically an epidemic of contagious disease that has extended across a vast region. It can affect the human populace or animal population. Common effects include health and/or the disruption of essential services that lead to economic or social costs. Additionally, it could be an uncommon or unplanned increase in the amount of cases around an infectious illness which already persists in a given region or populace, or may also mean the emergence of a considerable amount of incidents of an infectious disorder in an area, or population that's typically free from the infection. Additionally, pandemic emergencies may happen as a result of either manmade or natural disasters. They include diseases such as Ebola, Zika, Cholera, Avian Flu, Malaria, Yellow Fever, and Dengue Fever.
- **Sudden-Onset Calamities:** Just as the name suggests, these calamities happen swiftly and typically without prior warning. Normally, floods are a frequent kind of natural disaster related with abrupt migration of massive populations and food scarcities. Sudden-onset disasters normally happen with greater frequently in Latin America, Asia, and Caribbean countries instead of Africa. Once these disasters happen, they regularly cause multiple deaths and casualties. Additionally, earthquakes cause the highest amount of deaths and increasing infrastructural destruction. Communities which are at greatest risk of these kinds of disasters must recognize and reply to threats modeled by local weather structures, including the shape and outline of the land (Paul, 2011).

- **Slow-Onset Calamities:** The opposite is slow-onset calamities which comprise of droughts, environmental degradation, famine, deforestation, desertification, and pest infestation. The disasters are normally as a result of harsh weather patterns combined with bad land use. Conventionally, African communities, especially the poor, are the ones who have been exposed to these kinds of disasters the most due to poverty and social inequalities, environmental de-gradation from bad land use and quick population growth.

Additionally, slow-onset disasters are easy to prevent since they occur over a long duration of time, plus human decisions further contribute to these problems (or cause) them. Primary warning systems may be easily set up to diminish or even avoid the disaster.

1.4. PHASES OF DISASTERS

Generally, there is a predictable life-cycle that most disasters follow. Knowing about these life-cycle can help you plan in advance so that you don't suffer the impact as much. However, to a certain extent the best mitigation is adequate disaster planning, which reduces the impact.

Phases in Disaster Planning:

- **Phase 1: Pre-Disaster:** The first stage is pre-disaster. It involves realizing that a danger exists in the first place, and making mitigation plans early enough. The amount of support for these efforts depends basically on duration of time it's taken before the last major disaster of that kind in the region (Dufresne, 2019). It's also closely related to the Alert phase. Whenever previous warning is available, it comes with a fuss of preparations. There may further be debate on the reality and scale of the threat. (The phase may last for just some few minutes or even days, based on circumstances. Besides, it doesn't occur at all for earth-quakes or other unexpected impacts).
- **Phase 2: Impact:** For the Impact phase, there's an abrupt stunned silence. Here, usually the local media begin sorting out newsrooms and arranging their staff. External media are also scrambling for evaluations of the impact; however, the full impact is yet to be known (Dufresne, 2019).

- **Phase 3: Heroic:** This is followed by the Heroic phase. It involves mobilization, rescuing, and instant relief. There also lots of red-lights and sirens, high adrenaline, people helping others and so on. The media interest is also at its highest.
- **Phase 4: Honeymoon:** This phase occurs during the initial few days following the disaster. Typically, it carries along with it some great amount of public support as well as media coverage. It's also the time when first response team receives the greatest praise from public officers. Fundraising and volunteer recruiting are simplest during the honeymoon phase.
- **Phase 5: Disillusionment:** The succeeding disaster phase is Disillusionment. Once the immediate response has passed. Cleanup exercise commences and a sense of loss also comes in. Feelings of anger, blame, and depression are common at this stage.
- **Phase 6: Reconstruction:** It lays the groundwork for Recovery or Reconstruction that produces visible outcomes. Henceforth, victims learn how to form a new normal. Some even find a "silver lining" amidst all the havoc. Majority however suffer from prolonged social, psychological, and economic losses, (which become more noticeable during anniversaries, whereby victims and responders, counting the media, may encounter emotional reverberations known as 'anniversary reactions.')

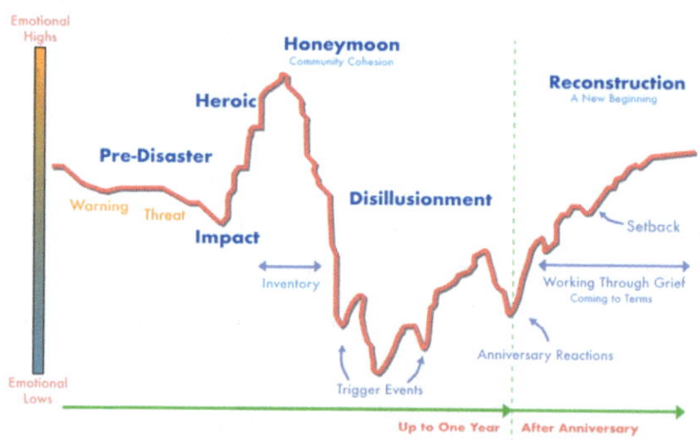

Figure: De Wolfe (2000). Training Manual for Mental Health and Human Service Workers in Major Disasters.

Source: https://files.eric.ed.gov/fulltext/ED459383.pdf.

1.4.1. Pre-Emergency Phase

Also known as pre-disaster stage, it's marked by fear and indecision. The exact reactions a community undergoes will depend on the kind of disaster that has hit. Those without any previous warning signs can trigger feelings of susceptibility, lack of security and fears of the future (Dufresne, 2019).

Conversely, disasters with prior warning may cause feelings of guilt, and self-blame for failing to take heed of the warnings. Pre-disaster stage can be as brief as a few hours or minutes, like during terrorist attacks, or might take several months, like during hurricane period.

The duration before a disaster strikes can be applied to evaluate how repeatedly a specific community has been exposed to various risks (risk mapping), including how well they are prepared. Then, protective measures may be applied depending on the disaster specific warning and accessible resources, example, identifying temporary shelters, gathering basic supplies, monitoring trends and planning evacuation routes (Dufresne, 2019).

- **Pre-Disaster Recovery Planning:** It provides an opportunity for populations to consider how they'll manage significant recovery matters, such as how to ensure government and vital services are up and running, plus where to find temporary shelter, rebuilding measures, and how to reestablish vital economic activity. Communities may, and should, actively take steps before getting impacted by the disaster to guarantee that the aftermath won't become a tragic and chaotic case by itself (Smith, 2003).

There are various kinds of community plans accessible to regional governments, which are especially useful for helping level out the road for post-calamity response and recovery. Furthermore, pre-disaster recovery strategies help to improve the recovering capacity of communities, plus their respective governments by developing an organizational structure for decisions, including actions that are to be considered post-disaster.

This process is normally wide-based and inclusive, building upon existent community plans or aligning with other disaster plans and policies. Developing a plan further helps populations to understand their existent resources and needs, identify opportunities for longstanding hazard avoidance and resilience building, plus establish relationships and procedures to ensure a quicker and more efficient recovery process in case of a disaster (Smith, 2003).

Remarkably, pre-disaster recovering plans recognize leadership as a driver for the recovery procedure, including the main responsibilities for regional government agencies and personnel. Based on studies from American Planning Association (APA), there exists several benefits of pre-disaster recovery plan some of which are; faster and highly efficient calamity recovery, establishment of defined leadership and decision-making systems for post-disaster recovery and enhanced ability to access necessary recovery resources; including improved community involvement in disaster recovery as well as community resilience efforts (Smith, 2003).

Furthermore, continuity of operations plans (COOP) may be established to ensure citizens don't experience considerable service disruption during and after moments of emergencies or disasters. COOPs is defined as the efforts within individual managerial departments or agencies to ensure primary (PMEFs) Mission Essential Functions continue being performed in a wide variety of emergencies, covering localized processes of nature, mishaps, and technological or attack-based emergencies.

Continuity of Government (COG) plans also work similarly to COOP, even though their main focus is establishing defined processes for allowing the state entity to continue with its essential operations after a catastrophic occurrence. COG plans establish processes for conserving facilities, apparatus, and records. Oftentimes, the COG plan forms part of a detailed COOP protocol (Smith, 2003).

- **Characteristics:** For many townships, cities, and counties, pre-disaster recovery, planning constitutes an emergent best-practice having a goal of establishing resilience to hazards, particularly since a number of these particular communities are facing similar hazards on a continuous basis. There aren't any established rules or templates that define the creation or execution of pre-disaster recovery strategies, and communities have established their plans in various ways. Among the most effective plans, share various common characteristics (Coppola, 2010).

Firstly, they have a holistic structure. While every community's pre-disaster recovering plan may appear different based on their requirements and priorities, furthermore it must consider the whole range of possible recovery decisions and activities which will be required in case of a disaster.

Some topics that can be addressed in this plan may comprise, business continuation and economic improvement, home repair and reconstruction, infrastructural restoration and alleviation, short-term recovery plans affecting long-term redevelopment, environmental renewal, and financial concerns among other issues of concern.

Based on the community's overall needs and priorities, certain programs like land-use planning tools can be passed pre-disaster to help make post-disaster recovery easier. Secondly, pre-disaster recovery plans must conform with other communal plans, such as all-inclusive plans, hazard avoidance plan and climate plans. Besides, many of these plans have links to state or federal disaster recovering frameworks, such as Federal Emergency Management Agency (FEMA's) Nationwide Disaster Recovery Structure (Coppola, 2010).

In some regions, The Framework is structured along the same model and within the same functional zone like the federal recovering support functions (RSFs), such as Community Planning or Capacity Building. Forming these linkages during pre-disaster recovery plans will allow people to better synchronize with government and federal associates during response, recovery, and also better leveraging external resources which shall become accessible to them.

Furthermore, pre-disaster planning needs broad-based involvement by local leaders and government officials. Whereas different offices may take a lead in forming a pre-disaster plan, the procedure itself needs active participation by different participants such as planners, key departmental employees, emergency managers, community leaders, and town/city managers.

Besides, a good pre-disaster recovering plan can make links to local or county-level players or resources that may provide extra capacity to the community post-disaster. Also, these policies must involve considerable community and stakeholder involvement and outreach.

Engaging community members and other stakeholders permits recovery planners to comprehend the existing (though sometimes contradictory) regulation conditions and variances between communities and populations in their dominion (Coppola, 2010).

Leveraging existing planning processes and systems, like hazard mitigation policy updates, resiliency procedures and adaptation tactics, are effective opportunities for engaging members of the public towards pre-disaster recovery plans. In Maui, for instance, the County government performed an extensive community involvement process which helped to shape seven other county-wide commendations at the center of their reconstruction strategies, plus addressing extra community-specific requirements and considerations.

Ultimately, pre-disaster recovery procedures must be flexible as well as scalable enough to meet the community's unique needs across various kinds of disasters plus recovery processes. In the case of Maui County, for instance, reconstruction procedures and protocols have set forth seven guidelines for recovery which don't have similar force of law like a rule or guideline; however is a flexible way to notify decision makers, establish policy, and apply a planning model in the wake of uncertainty (Coppola, 2010).

This covers a streamlined group of recovery actions which are flexible enough to meet a wide variety of hazards, which Maui may possibly face while also detailed enough to fast-track the reconstruction process. Even though pre-disaster planning on post-disaster recovery can be regarded an emergent best practice, still there are various useful guides accessible to local communities.

Pre-disaster planning for government bodies describe key recovery models and defines community based planning procedures, which align with National Disaster Recovering Frameworks. Besides, Community Recovery Management programs further offer guidance, tools, and even training for post-disaster recovering plan.

Furthermore, the U.S. Planning Association offers various essential reports as well as briefing papers which define both pre and post disaster recovery plan processes, apart from highlighting effective case studies from around the U.S. In various communities, both COGs and COOPs have been developed by Emergency Management groups; however, the planner plays crucial role still in alleviating disaster. For instance, planners may help in establishing continuity of operation protocols for Planning Departments to be added in the COOP (Coppola, 2010).

1.4.2. Impact and Flight Stage

When a disaster hits, it can trigger the relocation of huge numbers of people away from their houses. The amount of time this stage lasts will depend on the kind of disaster, amount of people affected, plus the distance to travel before finding sanctuary. Likewise, search, and rescue among other assistances, example, offering transportation, shelter or essential supplies to affected individuals may be performed to minimize loss of lives (Coppola, 2010).

The impact stage is also characterized by various intense emotional responses. Like with the pre-disaster stage, the particular reactions also rely on the kind of disaster that's happening. Slower, low-threat disasters also have psychological impacts that are unique from those of swift, dangerous disasters. Therefore, these reactions may vary from shock up to overt panic. Early confusion and disbelief normally are followed by the focus on self-preservation as well as family protection. Besides, the impact stage is typically the briefest of the six stages of disaster.

1.4.2.1. Risk Factors of Disaster Impact

Each year, millions of individuals become affected by human-based and natural calamities. These disasters may include explosions, floods, hurricanes, earthquakes, tornados, and fires. For such calamities, the risk of physical injury and death are real. You might also lose your house, belongings, and community. These stressors put you at the risk for both emotional and bodily health problems.

Stressful reactions following a disaster look pretty much like the typical reactions seen following any kind of trauma. Calamities can cause a wide range of cognitive and physical responses. You may further react to problems which happen after the event, including possible triggers and reminders of a trauma. Various factors make it highly likely that someone may have more intense or longer-lasting stressful reactions following a disaster. They include:

- **Severity of Exposure:** The rate of exposure to calamity is vastly related to the risk of future psychological troubles. At greatest risk are persons who go through these disasters personally. Coming next are persons in close contact to the victims. While at the lowest risk of long-term impact are individuals who simply had indirect exposure, like hearing news of the immense damage. Possible injury and life threats are key factors that lead many victims to psychological health problems. Scientists have looked at various severe natural disasters, like the Armenian earthquake, Mexican mudslides, and Hurricane Andrew which ravaged the US. Their findings reveal that around half of the survivors have suffered from distress and psychological health problems which require clinical care.
- **Gender and Family:** Nearly every time, women, and girls suffer the most negative effects compared to men or boys following

an impact. Disaster recovery is also quite stressful when kids are present in their homes. Ladies with spouses furthermore experience more agony during recovery. Caring for a family member at home who's extremely distraught is related to added stress for everybody. Additionally, marital stress is likely to increase following these disasters. Furthermore, conflicts between members of the family and lack of support at home may make it difficult to recover from calamities.

- **Age:** Adults within the age group of 40 to 60 years are highly likely to become distressed following disasters. Scientists thinking are that those within this age range have greater demands from both job and family. Besides, research on how kids respond to natural calamities is limited. Generally, kids show more intense distress following disasters than do grownups. Increased stress in parents is associated to worsening recovery in children. There are different factors related to the survivor's background, plus available resources which are essential for recovering from the disaster. Besides, recovery may be worse if you: weren't managing well before the calamity, dealing with other underlying stressors, or have low self-esteem. Other factors that could also lead to worse outcomes are; injury to oneself or other family members, bereavement, excessive loss of property, displacement, and separation from family.
- **Developing Countries:** These regions are at a higher risk when disaster hits since they are usually not prepared for such things. Besides, disasters in developing regions have highly severe psychological health impact, as compared do disasters that happen in developed nations. The same is true even for less serious calamities. For instance, natural disasters normally are believed to be less severe than human-caused. For developing countries, nevertheless, natural calamities have more serious effects compared to human-caused tragedies in developed nations.
- **Minimal Social Support:** Support from others is both a risk factor and resilience factor, depending on how it's provided. Generally, social support wanes following a disaster. This could be because of stress or the need for support network members to carry on with their livelihoods. Additionally, sometimes the responses coming from others that you depend on for support

may be negative. For instance, somebody may not take seriously your personal problems, needs, and pain, or even expect that you recover much faster than is possible. This is closely linked to longstanding anguish in trauma survivors. Following a mass trauma event, social conflicts, including those which have been resolved, might again re-emerge. Racial, ethnic, social, religious, and tribal divisions might further recur as individuals try to gain control to much-required resources. For families, conflicts can rise when family-members go through different experiences in the disaster. Ultimately, this establishes different paths of recovery which normally aren't well understood among fellow family members. People may also act as distressing reminders of the disaster towards each other, thus slowing down the healing and recovery process. Have in mind that whereas millions of people may directly be impacted by disasters, majority eventually recover. Humans naturally are resilient and have the capacity to heal from a disaster. Besides, people sometimes record positive changes even following a calamity. They may re-consider what's really important and start to appreciate what they value most in life.

1.4.3. Heroic Phase

Also known as acute phase, this stage is marked by increased activity levels and low rates of productivity. In this stage, there's also some sense of altruism, besides many community members display adrenaline-induced rescue action. Consequently, risk assessment can be affected. The heroic stage normally passes quickly as it progresses into phase.

The stage begins immediately following disaster impact and is marked by extreme, often reactive measures by humanitarian agencies who are responding to media news on high death rates (which could be 5 to 60 times the usual death rate). Generally, the priority is keeping affected persons alive. Security can be a vital concern in multifaceted emergency situations. Besides, disaster survivors aren't always conscious of their actions associated with adverse impacts of the calamity. Meaning they need help with identifying family members or resources that may have been lost in the confusion (Coppola, 2010; Substance Abuse and Mental Health Services Administration, 2020).

Furthermore, crucial services like provision of food, basic healthcare, sanitation, water, and protecting locals from violence or harassment must be rapidly organized. The stage comes to a close when estimate death rate drops below 1/10,000 individuals per day (on average, the crude mortality ratio for a stable population in the African continent is roughly 0.5/10,000 people per day).

- **Social Support:** Proper social support is a key factor towards recovery following any trauma, like a disaster. This social support improves well-being and reduces distress following mass trauma events. Getting connected to others furthermore makes it simpler to obtain knowledge necessary for disaster recovery. With social support, a victim can also get practical help in solving problems, some sense of getting understood or accepted and coping tips.
- **Coping Confidence:** Over and over, research has found that coping self-efficacy-"believing that you can do it"-is related to better mental health outcomes for disaster survivors. When you think that you can cope no matter what happens to you, you tend to do better after a disaster. It is not so much feeling like you can handle things in general. Rather, it believes you can cope with the results of a disaster that has been found to help survivors to recover.
- **Hope:** Better outcomes following disasters and mass trauma are possible if you have these particular attributes; optimism (since you can always hope for better in the future), confidence in predicting your life and personal ambitions or practical resources such as a house, job or money.

Open communication is crucial in times of disaster. Respondents may create a distinct department to manage crisis coordination, or use existing structures for this purpose. A particular site or venue must be designed in order to manage the crisis from a communication standpoint. While local media stations may help in disseminating information, sometimes they may not be enough and community members must come up with their own communication systems (Coppola, 2010).

The disaster team should be able to project the forms of information that would be needed. For example, they can utilize web-based platforms like social media to keep in touch with the local community in real time. Certainly, for some situations not adding information to the web platform would be strategic. It all depends on the nature and magnitude of the disaster in question.

Additionally, sometimes an organization might not want to publish a given disaster by putting information on the website. This assumes that the crisis is negligible and that stakeholders may be unlikely to hear any news about it coming from a different source. But in the current traditional and digital media environment, it's a misplaced if not risky assumption to say the least.

For mass notifications, contact details such as (emails and phones numbers) can be used to update close relatives of victims about their recovery progress. These contacts can also be any group which may be affected by a disaster, such as community members staying near a facility that had been damaged like dam. Mass notification systems offer a mechanism for individuals to respond quickly to messages, so that they can avoid the negative impacts of a disaster or get help swiftly (Coppola, 2010).

1.4.4. Honeymoon Phase

This stage is also commonly known as the Post-emergency phase. Usually, people draw together so as to solve issues in an intense show of community. Equally, media interest is profound during this stage. The phase normally lasts from between 1 week to 6 months. People's movements normally slow down. Therefore, allowing crucial services to be well established and maintained. In this phase, the ultimate cause of displacement might also be evolving (Babar, 2007).

As for the displaced persons, they are waiting for "something positive" to happen, so they can continue with their normal lives or possibly return home, assimilate with the native community or move elsewhere. When international support wanes, the emphasis moves to providing regional capacity of plus supporting community participation, mental healthcare, tracing reproductive healthcare and other similar services might also be initiated, depending on the available epidemiological data.

It's hard to set specific time restrictions on post-disaster time stages, or accurately describe the limits of every, even for a particular type of calamity. For instance, during the emergency stage of a hurricane event or flooding, it may just be some few days, or even as much as a week. Droughts and the resultant famine may last for several months or up to a full year or more (Babar, 2007).

This stage is normally marked by a drastic shift in emotion. In the honeymoon phase, victim assistance is easily available. Community bonding also happens. Optimism is high that everything shall get back to normal swiftly. Therefore, various opportunities are ready for providers as well as organizations to develop and build strong rapport with affected persons and groups, plus for them to foster interactions with stakeholders.

For health interventions, they usually evolve as capacity also grows. When the initial phase of the disaster has passed, a bit more comprehensive solutions can be provided. Post-emergency stage provides the opportunity to increase service delivery, improve quality and accessibility of services, plus to improve guidelines pertaining to addressing health matters.

Health stakeholders recommend certain actions that should be taken during the post-emergency stage. For instance, there's need to increase medical care for the survivors, create procedures for medico-legal collection of evidence, incorporate medical supervision into health-system structures and so on.

Upon understanding all the above stages, you won't then be stunned when during the honeymoon stage of disaster, community members may be originally quite appreciative of any help and responsiveness that you offer, but in other phases like disillusionment, they may react rather differently-apportioning blame and generally becoming rather critical (Babar, 2007).

If a particular company or organization is responsible for causing the disaster, then any progress made to address the matter during the honeymoon stage, or first to sixth week, shall be crucial in determining the help provided during the disillusionment stage, which follows the honeymoon one. It can be argued that the communication speed during the initial week is crucial to comprehensive communication, including the formation of a good plan, whereas the remaining 5 weeks are essential to positive delivery of the given plan, or for the least part a considerable start to it (Figure 1.3).

Figure 1.3: UN officials supplying essential commodities to a disaster-hit region.

Source: https://www.unocha.org/our-work/coordination/un-disaster-assessment-and-coordination-undac.

Internationally, the UN's office for coordination of humanitarian affairs (OCHA) is accountable for assembling humanitarian actors to guarantee a logical response to emergencies which need a global response. OCHA has a role to play in operations coordination during emergency situations. This may include evaluating situations and needs; approving common priorities; creating common strategies for addressing issues like discussing access, mobile funding and various other resources; expounding on consistent public messaging, and checking progress (United Nations, 2010).

In America, it's the Federal Emergency Managing Agency (FEMA) that coordinates federal operations and logistical disaster reaction capability required to save or sustain lives, reduces suffering, and secures belongings in a timely and effective way in communities which become overburdened by disasters.

Additionally, the CDC provides information for various kinds of emergencies, like disease outbreaks, severe climatic patterns, natural disasters, and chemical or radiation accidents. Furthermore, Emergency Preparedness and Response Programs are available for addressing responder safety as well as health during the responder and recovery procedures (United Nations, 2010).

As for UK, the organization responsible for providing communications calamity response is RAYNET. But they also have another agency called 4×4

Response that provides disaster response for off-road vehicles. In Canada, they have GlobalMedic which was set up in 1998 to serve as a non-sectarian and humanitarian-aid NGO for providing disaster relief solutions to large-scale catastrophes globally.

A few years ago, Time magazine gave recognition to the works of GlobalMedic by mentioning them in their 2010 issue. The agency has a roster consisting of well over 1,000 volunteers coming from all across Canada, including professional rescuers, firefighters, paramedics, and police officers. Together they give their time to appropriately respond to global disasters. Their workers are divided into the rapid response teams (RRTs) which run rescue units, water purification units (WPUs) developed to offer safe drinking water for those affected by the disaster, and emergency medical units (EMUs) that utilize inflatable field hospitals for providing emergency medical treatment (United Nations, 2010).

1.4.5. Repatriation Phase

Following an emergency situation, displaced persons are expected to go back to their area of origin, either individually or with assistance from relief agencies. Besides, repatriation can be either forceful or voluntary depending on how cooperative the victims are (Paul, 2011) (Figure 1.4).

Figure 1.4: War refugees receiving relief food and supplies from a donor.

Source: https://www.unhcr.org/voluntary-repatriation-49c3646cfe.html.

This stage also comes with a lot of disillusionment phase. Basically, communities, and people realize the bounds of possible disaster assistance. Any previous optimism soon turns to discouragement plus stress continues taking a toll on victims, not to mention negative reactions, like physical exhaustion and substance use may start to surface.

The widening gap between needs and help may cause feelings of abandonment. Particularly as the bigger community returns back to business as normal, there could be a heightened demand for services, since people and communities become willing to accept support. This disillusionment can last anywhere between a few months to several years, it's usually extended by a single or multiple trigger events, normally including the disaster's anniversary.

For victims, returning back home may not just fulfill the wishes of those who were once forced to escape, but also largely constitutes a positive effect of an abnormal case brought about by involuntary displacement. In addition, return, when done in strict observance to the law of non-refoulement, can be an essential component in ensuring the accessibility of asylum for people who may still want to revert to it.

When systematically executed, repatriation can discourage fake claimants from going through asylum channels. Repatriation, nevertheless, is a complex system, intertwined with the thorny matters of security, peace, reconciliation, and human rights; therefore, care should be exercised so as to really provide a sustainable and durable solution. Open policies and guidelines may thus be helpful in tackling this matter (Paul, 2011).

The unlimited presence of prohibited asylum seekers or individuals who don't qualify for global protection services comprises one of the biggest challenges faced by countries, plus the international community, in their endeavor to create an effective migration structure in full regard of fair asylum processes. The condition prevailing in many transition and destination states may undermine the entire concept of asylum, may pose a costly financial load on society plus be harmful to internal and, occasionally, regional stability.

One appropriate response is establishing a coordinated structure between humanitarian and executive actors to guarantee, on one hand, that people who need security continue obtaining it and, furthermore, that individuals in an irregular circumstance, who don't qualify, may be helped to return.

The approach can create situations for a foreseeable and durable structure, which is in accordance to respect of human dignity as well as well-being of victims, whose interests should be met in a positive way. By engaging in repatriation programs, help organizations like Red Cross can depend on certain distinct benefits afforded to them based on specific mandates and experience.

Through their unique universal network, these agencies are particularly well-placed to offer assistance during every phase of the repatriation procedure. Their existence, both in nations of asylum as well as countries of origin, plus their ability to mobilize regional branches in times of disaster represent a substantial strength and allows them to cover regions in a way that few other agencies can (Babar, 2007).

In many cases, whenever a large-scale immigrant exodus takes places following a disaster with dramatic suddenness, humanitarian bodies, with their historic impartiality, would act long before any intensive intergovernmental action is undertaken. Another unique advantage is the organized cooperation and support which exists between Nationwide Societies, plus the help they can get from International Relief Societies.

Humanitarian agencies in charge of repatriation also have well-founded traditions of shared values and principles, which provides them with the charitable mission to help and act forcefully in favor of displaced persons, returnees, and even refugees. Likewise, National Societies pursue various complementary activities that when joined with their expertise in the area of social or psychological counseling, and their special abilities in the human tracing arena, lend additional support towards getting involved in repatriation programs.

Every National Relief Society has the right to assess existing situations before taking a stand plus it's always mainly responsible for making the decision on whether to be involved in the repatriation program. Contrariwise, National Societies are in charge of decisions made and duty-bound to challenge efforts that seek to induce returns through instituting pressure (for example, withdrawing assistance to people who still needs global protection). For some situations, the condition to which displaced persons go back to might not be very stable. During such conditions, National Societies must assess if their resources can be better focused either towards informing returnees concerning potential alternatives to repatriation, or the consequences of making such a move (Babar, 2007).

- **Fundamental Principles in Repatriation:** Relief agencies consider of importance certain fundamental principles when helping people in disaster areas. Apart from showing humanity, independence, fairness, and neutrality, they should also provide these working principles as much as possible.

- Operate in the beneficiary's best interest.
- Freedom of choice: support is offered with respect to each person's free will to avail themselves of the services or not.
- Confidentiality: Individual data and other related information about the beneficiary is disclosed to 3rd parties only when the person provides their consent.
- Free service: Absolutely no charge is imposed; and
- Free decision on whether to become involved in repatriation-based activities. Relief agencies are also mandated to respect applicable global human rights and principles pertaining to individuals who have mentioned they need international protection.

Safe repatriation, with respect and possibility of future empowerment programs backed by relief societies should be done in full regard to the repatriates' protection as captured by the UNHCR guidebook on voluntary repatriation. Any favorable decision in regards to safety which has been arrive at before the real repatriation, is normally subject to re-assessment in regards to objective circumstances which, consequently, suggests acting on precise and verifiable data about the receiving nation.

This could possibly lead a relief agency carrying out its own evaluations, or coordinating its decisions alongside government agencies and other relevant NGOs. The fundamental premise should be the global right for a person to return to their country plus that no circumstances may be imposed that limit the returnee's central rights, or that aren't at the same instance enforced on local populations (Babar, 2007).

- **Phases of Repatriation:** Repatriation can be subdivided into three key phases. Based on their role and level of expertise, relief societies can be included in all three stages. It's essential that state bodies offer rich details on the repatriation procedure, including what repatriation programs will be involved in concrete terms. Whereas it's generally the role of government to design and follow unambiguous and defined repatriation plans, considering the costs involved, it's also the government that must cover the most costs.
- **Phase 1:** It's the duration just before actual repatriation project. It involves preparation as well as reorientation in the new

settlement area, which is usually away from the disaster zone. Advising for victims about to be repatriated is also important; they need to be acquainted to fundamental principles, methods, and objectives of the process. This will also help them form their own independent judgment, regarding their willingness to carry on with the process or otherwise. Relief agencies must also undertake info-finding missions to identify potential projects for networking with other respective agencies; it's also important to make thorough assessments, planning, preliminary installations and appropriate maintenance of convenient communications structures connecting capitals, victims' areas of residence within the host nation and regions of return; plus use the tracing system to develop contacts between repatriates, together with their kin or friends coming from the same area or beyond. Some issues that can be covered during repatriation agreement are: conditions of the victims, individual rights and status, documentation, asset transfer, access to property and land, plus operational issues consisting of (coordination mechanism, security responsibility, logistical arrangements, condition of aid agency workforce and property, in addition to communication systems) (Emerald Group Publishing Limited, 2015).

- **Phase 2:** This is where psychological counseling for repatriated victims begins. They need to be prepared for a new life outside what they were used to before the disaster, for some people this may not be an easy thing to do. The individual concerned must also be provided with precise and reliable information concerning, the safety condition in the receiving venue, socioeconomic status of the newly repatriated persons and available opportunities for make a living once more. Oftentimes, disaster leads to loss of livelihood for victims, their businesses are shattered and crucial infrastructures such as electricity grids also damaged. The rescued victims need to know that they can return back to their former activities, or possibly find new ways of catering to their daily needs. Preferably, there should be opportunities for skills development, training, and education for learning new concepts, or brushing up on knowledge already acquired previously before the disaster happened. The prospects of financial support must also be considered since displaced persons are usually disadvantaged economically. They will need funds to build new homes where necessary, replace essential items lost to the disaster such as clothing and so on. Other needs that repatriated persons have included; essential medical requirements, sufficient transport and food, plus protection during the whole process, for instance, by providing security accompaniment to returnees (Babar, 2007).

- **Phase 3:** Reception and reintegration into the new society following repatriation should be done in a timely and orderly manner. For situations where the disaster was caused by war, this is especially important in contributing to peace-building as well as helping to prevent resurgence of conflicts and displacement upon their return.

Besides, a long-term development perception must be firmly integrated into the repatriation project for convenience purposes. The relief team must undertake its regular activities without any political pressure from outside, since sometimes politicians may want to interfere with such processes to popularize themselves. The repatriation process serves the interests, rather in a balanced way, of the repatriate as well as community to which they are returning; but it's not the case when repatriates are systematically provided preference over regional populations, or if they're systematically victimized against (Babar, 2007).

Furthermore, special attention must be provided during the distribution of relief plus support, and towards achieving self-reliance through promoting integration measures into the socioeconomic life of the area of origin. Not to mention reconstruction and reintegration actions must be supported and applied according to a really non-discriminatory and neutral approach, which targets social matters alongside vulnerable returnees. Agencies must also favor the procedure of reconciliation through supporting confidence-building programs aimed at reconciliation (example, training in conflict dissolution, peace education and human rights, ladies' involvement in the preparation and implementation of programs (Srinivas, 2015).

1.4.6. Reconstruction Phase

This stage is marked by a general feeling of recovery. People and communities start to assume obligation for reconstructing their lives, and individuals also adjust to an all new "normal" whereas continuing to nurse previous losses. The re-construction stage often starts around the disaster anniversary and might continue for a while beyond that. After catastrophic events, the reconstruction stage may last for a couple of years (Birkmann, 2007) (Figure 1.5).

Figure 1.5: Community members reconstruct houses destroyed by a storm.

Source: https://www.ghba.org/rebuild-confidence-hire-professional/rebuilding/.

The principal goal of re-construction work is not just to reconstruct dwellings and infrastructure destroyed or compromised by the disaster, rather to lessen the future vulnerability, therefore making at-risk zones more economically safe, and enhancing safety plus life standards of the population.

Rehabilitation or reconstruction is an optimistic stage where the victims are actively picking up the pieces of their lives. Once permanent solutions have been found, the focus moves from mere relief to development. The goal is helping affected communities to become self-reliant.

Basically, the responsibility of offering assistance is given to affected communities, local authorities, developmental agencies plus other non-governmental establishments. Since humanitarian and development experts have unique approaches to offering assistance, plus the infrastructure needed for relief may be insufficient for development, it means the programs may eventually be re-oriented and restructured (United Nations, 2010).

The goal of the reconstruction stage is restoring the living situations of disaster-prone populations. Humanitarian professionals help the victims to overcome effects of war or natural disasters, by providing conditions fir lasting improvements which lead to improved housing, infrastructure, or income-generating prospects.

- **Humanitarian Aid:** Relief agencies have an extensive and solid experience with regards to the area of rehabilitation and reconstruction. Post-disaster rebuilding efforts cover both physical reconstruction and socioeconomic recovery of affected communities. Projects involve the restoration of public buildings for example (schools, roads, bridges, and hospitals) including damaged houses and rebuilding of water supply structures. Humanitarian aid also aims, in case conditions permit it, to reinstate local economic activity plus provide social protection to victims so as to allow them to fulfill their needs independently.

Furthermore, Humanitarian Aid seeks two key goals: to reinforce the skills and competences of aid recipients plus to ensure that its actions produce sustainable outcomes. The reconstruction stage is regularly an opportunity for mutual humanitarian aid, including the building of cooperation efforts (United Nations, 2010).

- **Context:** Following a disaster and certainly during normal periods-a person's living standard are determined by different factors like their housing situation, income, health, and quality of food which they eat. The living circumstances of vulnerable groups may be considerably enhanced by reinstating their access to primary necessities which are (water, hygiene, sanitation, arable land, and livestock) plus by fixing roads and bridges. Moreover, restoring income-creation activities provides the framework for much-necessary sociopolitical stability (Birkmann, 2007).

It's therefore remarkably important to guarantee that plans done at the start of a reconstruction stage be precise and attuned with needs expressed on the ground. Partnership with regional authorities during this phase is crucial to ensure the assistance given yields sustainable results.

Additionally, reconstruction activities need a significant coordination effort between regional and global humanitarian and development agencies. Preferably, donors, and partners must agree to subdivide tasks in order to make good use of all the available financial resources.

For both manmade and natural calamity situations, the effects can be alleviated to a huge extent through sufficient planning and preparedness. Harmful impacts of manmade tragedies may be managed, when socioecological and economic impacts of our activities are considered and growth decisions made consequently. Consequently, even though we can be sufficiently prepared for natural disasters, we just can't completely eliminate its impacts (United Nations, 2010).

- **Identifying the Problem:** For purposes of designing a response strategy which addresses sustainability matters, it's essential to understand the general causes and long-term effects of a tragic disaster. The problem-tree to this matter was built by Marcus Oxley from CARE Australia; a renowned expert in calamity emergency and relief activities.

Risky building practices which cause large-scale destruction are, in fact, the result of a fatal blend of lack of knowledge about safe building actions, lack of technological opportunities for safer building, plus a fatalistic attitude concerning the prospect of a disaster. They are typically coupled with misplaced possibilities that lead to extra cash spent on facades, including embellishments in homes than on safe building practices in economically well off homes; and buttressed by poverty which prioritizes the day-to-day meal over much safe shelter (United Nations, 2010).

Such anomalies remain at the core of why disasters have such a big toll on victims. Calamities like super cyclones and major earthquakes, particularly affect the poor in negative ways since the damage caused is immense. The effects are, certainly the most noticeable and instant, loss of life, infrastructure, and personal property. These long term and tough outcomes increase the victims' vulnerability to weather elements, loss of livelihoods, heightened poverty, economic decline, malnutrition, and increased social inequalities and strife.

- **Common Response Mechanisms:** Normally, post-disaster response is available at four levels which are as follows:
- Emergency relief: This is provided immediately following the calamity, normally lasting from anywhere between the initial 24 hours to roughly 2 to 3 months and attending to immediate housing, food, water, and healthcare assistance.
- Reconstruction immediately following relief and stretching to a duration of roughly two years, intended at rebuilding the primary physical infrastructure as well as shelter to allow people to start afresh; apart from,
- Rehabilitation, which looks at long-term efforts of reinstating lost daily occupations, introducing fresh economic opportunities and enhancing land and water managing processes in order to lessen

people's vulnerability and increase abilities to handle future disasters.

Readiness, it's a response that should preferably be a practical measure, whose mission is enhancing preparedness in recognized vulnerable areas by introducing systems and techniques of construction which mitigate effects of future disasters.

1.5. DISASTER: THE OPPORTUNITY

The best approach is not to look at disaster as a glass that's half empty but rather a half full one. Yes, it's a great loss but also an opportunity at the end of the tunnel to start all over "again." The opportunity to start a development process in a rather sustainable way. People have a chance to establish systems, technologies, and procedures that enhance quality of life, apart from being in-sync with local geo-environmental situations.

It's also possible to perform so at any given moment, plus at a massive enough scale not just because one has practically a new space to begin with, though also since people's mindsets concerning conventional (as practiced) structures of construction (plus other development systems) are altered in a stroke.

Some communities that would be highly conservative concerning any changes in their building norms, may question how they and their predecessors have been doing construction and seek to "improve" systems. This way they can easily comprehend the limitations of previous systems, plus be open to change that leads to improvement (Geipel, 2012).

1.5.1. A Holistic Strategy

The reconstruction process involves part-time or complete relocation, plus rebuilding the vital physical infrastructure and housing) so that exposure levels are minimized and families are capable of getting back on their feet. Reconstruction thus paves the means for long term rehabilitation.

Additionally, rehabilitation mostly addresses the emerging or increased poverty rates that have emerged because of the disaster. Careers and income generation processes in the construction industry provide a prompt and emergency boost for the local economy. It's followed by long-lasting improvements in land and aquatic system management, including economic opportunities that attempt to upscale local economies and lessen community susceptibility in a sustainable way.

For viable development to happen in continuance with re-construction, it's essential that the final objective isn't limited to just getting individuals back up to their base line points before the disaster. The intervention across a longer term must result in enhanced quality of life, including lesser levels of vulnerability. When families are finally tuned to recollecting the parts of their lives destroyed by the calamity, models of enhanced building procedures, sanitation, rainwater harvesting, sewerage, better land and water managing, etc., can be steadily introduced (Veenema, 2003).

Following a cyclone, for instance, development activities should focus on enhanced shelter, terrestrial, and water management plus livelihoods restoration. Loss of property as well as economic activity should be among the first things to be addressed. If these things are addressed, people may jump back to their usual routines very soon following the calamity.

However, when proper measures are not put in place, families may be unable to bounce back to daily work causing longer-term economic distress. It's a rather strong argument that favors "total rehabilitation" in comparison to just reconstruction. The latter and rehabilitation must be in a smooth scale with restoration efforts. The matter of correct timing plus speed is nevertheless, significant. A holistic system doesn't negate the requirement for a quick response to prompt re-construction.

Time and again, it's been observed that people will go back to their previous unsustainable activities very soon, in case timely inputs aren't made and structures that guarantee long term sustainability of material and skills development aren't established then it means people will continue suffering (Veenema, 2003).

Additionally, rural families are usually at a disadvantage after a disaster compared to their more urban counterparts. They aren't near amenities or resources that facilitate faster recovery from calamities, such as road networks, markets for goods, communication systems and so on. Besides, in many cases following an earthquake any house extension or new constructions are performed in a rather unsafe way, since neither materials nor the skills of enhanced building technologies are available.

1.6. GOING FROM RE-CONSTRUCTION TO REHABILITATION

The re-construction process of home and community infrastructure, generally, forms a crucial entry point for rehabilitation procedures. A reconstruction project is the primary step towards restoration and upgrading of local habitat. Furthermore, it introduces enhanced systems of building, establishes basic building component supply, develops the skills as well as management abilities of families, regional agencies and community artisans in a limited area, and develops local data and knowledge systems. These structures allow for "advanced building" (Weil, 2004).

The holistic approach of "Habitat" which links the procedure of housing together with the ability to make and apply informed choices touching on building development, habitat enhancement, and economic advancement is the greater objective. Re-establishing individuals' lives through positive rehabilitation efforts covers:

- Moving up the hierarchy from home to habitat and finally livelihood.
- Local awareness building, including training skills for everybody so that individuals gain proper control over the re-homing process.
- Capacity building and networking to business-livelihood support.
- Creating livelihood interventions for both farming and non-farming sectors, based on emerging economic opportunities for building economic surpluses (which can be redirected to responsive housing).
- Establishing a good foundation for community access and institutional housing finance.

1.6.1. Facilitating the Development of Sustainable Livelihoods

An effective response approach understands the requirement for having building materials, buildings, as well as livelihoods, plus catalyzing the adaptation of this particular need into demand. Therefore, demand for (sustainable) construction technologies and construction activities may be offered through sustainable enterprises.

The ideal response approach addresses the existing (immediate) need of re-construction through regional building technology-based programs. Reconstruction programs, when developed to include local man-power, offer the essential (although short term) jobs that lead to an instant jump in the area economy. Similarly, building material and skills-based local enterprises can ensure continual supply of top quality construction materials and skills. For the long term, this may probably cause a sustainable enhancement in shelter situations while also expanding livelihood options within the area (United Nations, 2010).

The re-construction project at the outset offers a major improvement to new enterprises. This forms the preliminary captive market, while also providing crucial visibility to emerging technologies and enhanced systems of building and also (when systematically approached) fosters the general acceptance of these fresh "products" within the marketplace.

A complex re-construction program will essentially involve an associated process of passing knowledge to affected populations on the attributes of safer construction; therefore, inculcating a gratefulness of the upgraded systems. Following the initial re-construction stage, families would ideally opt for these resources and methods to extend their homes.

An essential attribute here is that selected materials and techniques must match the paying ability of the targeted populations. A multi-pronged method is needed here: First, there's need for choosing of enhanced technologies and construction structures must bear in mind the general affordability of the victims (United Nations, 2010).

It involves accurate assortment of raw materials, manufacturing processes and delivery scales. The optimal blend of large industry-based materials, plus local enterprise based manufacturing with materials obtained from local building centers. In addition, a parallel intervention for enhancing quality-of-life through expanded livelihood options, plus enhanced land, water, and resource management activities that result in improved purchasing power in local communities.

Additionally, interventions for housing and livelihood financing are needed that allow people to access existing building options. Collaborations can be made with local NGOs to supply building material for construction work needed after disaster. These centers can also supply enhanced building materials, technology, and skills needed for continuing Restoration Programs.

For the long term, they're envisaged as venues for habitat management to the local community on matters like housing, sanitation, water storage, and domestic energy. They can be one-stop stores for all regional habitat needs, plus access to home financing for those who must rebuild their lives from scratch. Central production and service stations may also be developed, however these are normally structured to be partitioned into down-scaled construction material enterprises, becoming the nucleus of a great number of de-centralized production units that are spread across the reconstruction region (World Bank Group, 2017).

Furthermore, the skills and technology transference process which lies at the heart of home construction is already enabling building material production as well as construction for enhanced livelihoods. It's designed for introducing new skills as well as capacity for enhanced cyclone-resistant building structures within the regional area, in form of enterprises. The enterprises can continue building new homes for the destitute, or extend and upscale old homes long after the initial reconstruction interventions are complete.

Besides, project funds that may have been donated during the short-term may be applied to finance construction of houses, and establish material-based enterprises; plus, serve as an investment for livelihoods, furthermore capacity-building and data dissemination may pay dividends through way of:

- Empowered, informed communities;
- Improved building material supply;
- Enhanced economies;
- Reconstruction in the perspective of development.

The re-construction stage is more complex compared to the rehabilitation stage and needs a thorough planning process, which takes various factors into account, such as sectoral and developmental policy. Furthermore, costs usually tend to surpass national funding opportunities, especially for less developed nations, which should seek external financing. Generally, the idea of re-construction isn't just to repair infrastructure, however also to plan and implement policies and strategies intended to achieve (World Bank Group, 2017).

Long-term socioeconomic development through activities taken to minimize vulnerability to negative events, drive resources to works which anticipate the outcomes of natural phenomena, apply preventive measures, manage risk and create systems to offer forecasts and initial warnings. Some of these systems have already been recognized by WHO as the standard for rebuilding following a disaster.

They stress that quick mitigation procedures can minimize the physical and socioeconomic effects of calamities up to manageable standards, therefore contributing to longstanding development. Integrated systems that integrate the managerial and alleviation of natural risks into development, plus investments in project planning must also be encouraged. It's an affordable, multidisciplinary, and multi-sectoral technique that profits the most destitute and most vulnerable populaces, apart from encouraging development assistance organizations to consider assisting the disaster victims (Birkmann, 2007).

Other measures that can be of assistance include; integrating natural hazard considerations within the primary phases of assimilated development planning, plus investment project development. Not to mention, putting more importance on the risk reduction measures when assessing investment projects.

1.7. HEALTH SECTOR RE-CONSTRUCTION PLANNING

The impacts of disasters vary considerably. For instance, in the event of an earth-quake, the effect on the surrounding community may be so immense that the ability to offer health solutions is sharply lessened for a long period afterward. Likewise, floods, don't just impact the land as well as the production sector, however, also cause severe damage to existing health systems, and hospital equipment that typically causes functional breakdown of health facilities.

The re-construction duration presents both challenges and prospects in the wake of disasters. In regard to health amenities, reconstruction provides victims. The prospect to make structural and organizational modifications that reduce imminent susceptibility of health infrastructure, apart from increasing the effectiveness of particular services (Srinivas, 2015).

Renewal in the health-sector must take into consideration these following issues; Analysis of the amount of risk whereby the region affected by the calamity is exposed and its ensuing vulnerability to probable future events. Besides, better decisions may be done concerning whether to re-build the destroyed infrastructure in just about the same venue or choose an alternative site that's much safer.

Moreover, the re-construction stage provides an excellent chance to make positive changes in health services networks, and care model through avoiding development of advanced hospitals, or hospitals with massive amounts of beds. Besides, during the re-construction procedure, consider the basic-care-based model that facilitates response ability at the essential care level, apart from promoting an appropriately structured hospital network that's based on the models of referral as well as counter-referral.

The model and principles of 'safe hospitals' initiative must be rigorously executed when developing fresh health facilities. Guaranteeing that hospitals and every possible level of health facility meets structural, non-structural, and organizational safety needs. This work is usually the responsibility of nationwide health authorities, with involvement from technical specialists, who are in-charge of Health Emergency programs (United Nations, 2010).

The re-construction stage also provides an opportunity for strengthening disaster preparedness, plus mitigation processes and activities within the health sector. Besides, risk prevention and reduction must be an integral part of re-construction programs in the sector. It's also important to consider new technology developments so as to reinforce architectural models for fresh architecture and equipment, and/or updating remaining buildings and equipment. It's also important to develop and update accreditation processes to integrate the safe-hospitals initiative in disaster circumstances.

1.8. PREPARING FOR DISASTERS

Disasters, especially natural ones, can happen at just about any time. Some give previous warning signs like a storm that precedes flood. But others, like earthquakes, have little to no warning. When a disaster strikes, preparation time is completely gone. However, there are still certain measures you can take to lessen the negative effects of disasters.

The best strategy when coping with natural disasters is preparing by developing a solid plan before it hits. Disasters vary with regards to size, from small events like a storm affecting just one suburb, to massive events which can impact entire cities or massive areas which cut across state boundaries (Elliott and Hsu, 2016).

It's always good to have an insurance policy to mitigate any losses the disaster may bring, plus confirm that you have sufficient cover against any kind of disaster that you may be vulnerable to. Likewise, check the overall sum-insured to avoid a situation of being underinsured.

Additionally, consider taking a household inventory, plus store the necessary copies in safe locations away from the disaster area. By cataloging your belongings using a home inventory, it would be easier to recall all items of your residence in case you accidentally lost everything. Through preparing a home inventory, it can help in determining if you have sufficient insurance plus can save you time as well as unnecessary headaches when filing for a claim after the disaster strikes (Cutter, 2012).

It's important to safeguard your family, house, business, and properties from calamities. Insurers provide communities with the best way to achieve this. After a natural disaster, the insurance firms can pay up claims to policyholders who have been affected by the disaster, therefore allowing once devastated communities to rebuild and get back to normal.

- **Emergency Preparedness Kits:** An emergency kit is among the most essential priority tools for disaster preparedness. It includes all the basic products for family members in case of calamity. Additionally, scientific technological progress or nature itself can give warning to impending disastrous events and people should be ready always with basic supplies to preserve life. Some items found in an emergency preparedness kit include; portable, battery operated radio and additional batteries, whistle to call for help and first aid kits.
- **Safety Measures, Risk Assessments, and Mitigation Plans for Environmental Disasters:** Safety measures are processes used to warn and prepare people about impending disaster. Traditional community warning systems are telecast used for weather forecast. It may also include stocking on emergency kits such as a portable radio, spare batteries, dry food, matches, torches, and fuel lamps depending on the kind of disaster that's foreseen.

Disaster risk assessment, meanwhile, is a process used to determine the overall nature and extent that a risk carries, through analyzing hazards and assessing existing circumstances of vulnerability that mutually could possibly harm exposed individuals, property, services, livelihoods or the environment that they rely on.

Mitigation on the other hand are actions taken to avoid or diminish the risk to people's lives, property, socioeconomic activities and natural resources coming from natural hazards—this is key to preventing the negative effects of disasters. Some common mitigation practices include adoption of zoning, land-use policies, and building codes. These are needed to prevent and reduce any potential damage from hazards.

2

Factors Leading to Environmental Disasters

CONTENTS

2.1. INTRODUCTION

Environmental disasters have a negative impact on ecosystems. The environmental disasters usually take a short time, but their impact has a lasting effect especially on biological organisms and also the environment affecting their habitat. Sometimes these disasters change the physical environment and may have an irreversible impact to the ecosystem. In some cases, the degradation or impact can be corrected by environmental rehabilitation and restoration (Figure 2.1).

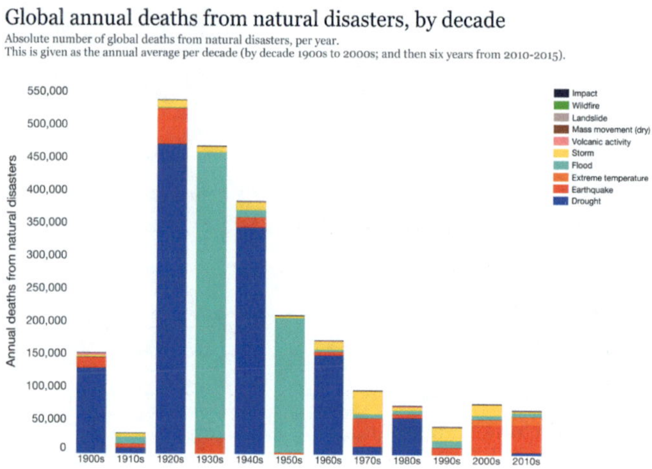

Figure 2.1: Average annual global deaths from natural disasters by decade, 1900–2015.

Source: https://ourworldindata.org/natural-disasters#number-of-deaths-from-natural-disasters.

Disasters can be categorized into different groups; environmental disasters can be as a result of natural causes such as changes in climate or weather patterns. Natural disasters can also be classified into five categories they include geological, hydrological, meteorological, wildfires, and space disasters. Geological disasters include floods, volcanic eruptions, and tsunamis. Meteorological disasters are blizzards, hailstorms, cyclonic storms, ice storms, heat waves, thunderstorms, droughts, and tornadoes. Even though the natural disasters do not directly involve human activities, the effects of the disasters can be made worse by people. A good example is the increased destruction from the tsunami of 2004 in the Indian Ocean in places where coral reefs had previously been damaged as a result of urban construction and development. In places with less damaged coral reefs or those that were healthy, the reef functioned as a buffer by deflecting the force of the waves. An increase in the occurrence and severity of natural disasters leave a large population of people every year especially those that are poor or marginalized (Babar, 2007).

The second category of disasters is those that are a result of human activities. Some of the examples of human-induced environmental disasters include oil spills, chemical spills, and nuclear incidents. Other human induced environmental disasters can be caused by wars and terrorist activities that can disastrous to the environment and ecosystem. In many cases, the environmental disasters caused as a result of human activities have a long lasting effect on the environment in comparison to those caused by natural events for example the oil spill in 1989 at prince William sound still has continuous environmental effects. It has been several years since the oil spill, there are still significant oil deposits which have proved to be toxic to many aquatic species and has persisted throughout the area.

Climate change as a result of human activities also increases the frequency and intensity of climate related disasters and these results in a high number of deaths and injuries as well as increased losses in property and in relation to the economy. Vulnerability is also accelerated by the ongoing environmental degradation and other factors such as the lack of effective policies by the government (Babar, 2007).

Despite environmental disasters having a negative impact on the environment and ecosystem, they also lead to an increase in attention towards the ecosystems that are threatened. In other cases, they lead to increased oversight by the governmental, intergovernmental, and nongovernmental organizations resulting in legislation, reducing the impact of environmental disasters in future. A good example is still the Exxon Valdez oil spill led to much stringent regulations for the shipping industry dealing with oil. Extra funds were also allocated to cleaning oil spills in the event they occur again. After the 2004, Indian Ocean Tsunami, the united nations organized a tsunami warning system to alert the residents in the event another Tsunami was to hit the shore (United Nations, 2010).

Overtime people have believed technology could solve and meet challenges caused by natural disasters, however as the earth's population continues to grow and there is an increased demand for natural resources;

the damage, depletion or degradation of these resources are more likely to cause environmental disasters in both intensity and number. Healthy ecosystems have some capacity to deal with disasters by reducing the risk of these disasters for example wetlands, forests, and coastal areas can reduce vulnerability to hazards by providing support to livelihoods and also functioning as physical barriers in the reduction of the impact caused by these disasters. Given that vulnerable populations are the most affected by environmental disasters and sometimes the most dependent on natural resources, it is necessary to find suitable ways to cushion them from the impact through restoring the ecosystem and also the sustainable management of resources. The above measures play an important role in population's ability to prevent, deal with, and recover in the event of a disaster. Although some of the disasters cannot be prevented, there has been a significant decline in the number of deaths recorded as a result of natural disasters from the historical data provided above. This is with the use of technology to make early predictions, infrastructure that is resilient, emergency preparedness, and response systems. These modern technologies provide tools that are useful in dealing with the problem of balancing human needs and equity of the environment.

This volume will discuss and explore different factors leading to environmental disasters, their impacts on the environment and the ecosystem. There are several factors that increase the likelihood of environmental disasters they can either be human, economic, natural, and social factors some of the include poverty, a high population density, rapid urbanization, environmental degradation, soil erosion, natural activities in the earth's crust, tectonic movement, moon activity, mining, deforestation, air pressure, global warming, pollution, and the lack of awareness (United Nations, 2010).

2.2. EXAMPLES OF ENVIRONMENTAL DISASTERS

- The dust bowl-in the period between 1930 and 1940, conditions of drought and farmland depletion led to severe dust storms that reached 10,000 feet in the sly and was conveniently referred to as the black blizzard. This resulted from the agricultural driven practices of farmers from the United States. The cultivated the virgin topsoil of the Great Plains, they practiced monoculture farming. The number of people displaced was approximately 2.5 million and also it was during the period of great depression adding more problems to the country.

- Poison in Minamata Bay-The Chisso Corporation in Japan between the periods of 1932 to 1968 released wastewater with mercury levels that were high into the sea around the city. The mercury poisoning marine life led to the poisoning of inhabitants who later fell ill and this led to the discovery of a neurological condition referred to as Chisso-Minamata disease. A number of 1700 deaths of people have been recorded to date. The disease causes convulsions, sight, and hearing loss, paralysis, coma, and finally death.
- Ecocide in Vietnam-military from the United States sprayed rainbow herbicides in the jungles of south Asia. This was between 1961 and 1971. They included agent blue, purple, and pink. Orange accounted for half of the amount that was showered over the jungles. This led to the loss of human life and it was termed as horrific. The large scale destruction of the environment led to the development of the word ecocide.
- Death in Bhopal-it is referred to as the world's largest catastrophic event. It occurred in the city of Bhopal in India.32 tons of chemical gases that were deadly leaked into the city and an estimated 9000 people immediately died as a result of air-borne poison that was invisible. The death toll that followed over the weeks was an estimated 20, 000 lives and other residents sustained injuries in thousands and this was permanent. Up to today, the site still remains a toxic waste site and acts as a contaminant to the ground water of Bhopal.
- The Chernobyl catastrophe-it started with a wind scale in 1957 and was followed by a three-mile island in 1979 but in 1986 in the Chernobyl nuclear plant in Ukraine the nuclear reactor had a melt down and this became the worst nuclear related disaster recorded in history. The total number of deaths recorded was 4,000 and this is according to the report by the Nations Chernobyl Forum Report (Figure 2.2).

Figure 2.2: Aftermath of the Chernobyl catastrophe.

Source: www.disasterium.com/10-worst-man-made-disasters-of-all-time/.

The Exxon Valdez oil spill-it is the most famous oil spill in the history of mankind so far it is not the largest record. In this catastrophe 11 million gallons was dumped in Alaska in the year 1989. 160 gallons of oil was dumped in the Persian Gulf in 1991 and was the largest record. The incident of Exxon Valdez was an eye opener and created public awareness of the consequences and this led to congress pass the pollution act in 1990. The effects are yet to completely clear up (Figure 2.3).

Figure 2.3: The oil spill of Exxon Valdez.

Source: www.disasterium.com/10-worst-man-made-disasters-of-all-time/.

- Dying oceans-in 1992 a large number of people lost their jobs when the cod population suffered in the waters of new found land. The effect on the marine system was devastating. Currently fish stocks are overfished from Iceland to Chile. This means the marine ecological systems were being over exploited and this was beyond their limits or capacity. The reducing populations of fish affect the entire ecosystem and this does not apply to only predators. They also affect the livelihoods and economies of human beings.
- Perfect storm over Lake Victoria-Lake Victoria is the largest lake in Africa and is a center for perfect storms leading to environmental crises. The environmental crises include chemical and pollution with sewage that is raw. It is also inhabited by hyacinth, a place for overfishing, and also the overgrowth of algae which suffocates plants and animals. In addition to this, the border of the lakes is also shrinking by about 150 feet in some areas. There are 40 million Africans who depend on Lake Victoria for their livelihoods and sustenance. These countries include Kenya, Uganda, and Tanzania. This makes them susceptible to environmental disasters.
- Destruction of the Amazon-A percentage of the forest cover at the Amazon rainforest has been destroyed as a result of tree logging and other human activities such as cattle ranching, soy farming and the construction of roads. Forest biodiversity is also damaged and this is difficult to estimate. The large amounts of carbon released are later held by the plants and trees of the forest and this can contribute to the process of global warming. Scientists have suggested different measures to mitigate the effect and this includes job creation and sustainable development (Figure 2.4).

Figure 2.4: Deforestation in the Amazon.

Source: https://www.nationofchange.org/2018/11/27/amazon-rainforest-deforestation-at-its-worst-equivalent-to-the-size-of-987500-soccer-fields/.

- Global warming: global warming has led to jelly fish swarms, melting of the polar ice caps, melting glaciers, desertification of wetlands and the spread of pest and diseases. The effects of global warming are caused by an increase of atmospheric greenhouse gases. In 1963, the first climate change conference was held as a result of the increasing urgency of the matter. With industrialization and over exploitation of natural resources, the effect is reaching a threshold where we can no longer be able to influence climate change.

Some of the factors leading to environmental disasters are discussed in subsections.

2.3. POPULATION GROWTH

There is a relationship between the rapid population growth and an increase in losses from disasters. If there are more people in the place where disaster strikes, there is likely to be more impact. With the growing population, it is inevitable that more people will be affected by the disasters because more people will be forced to live and work in areas that are unsafe. With the growing population, people also compete for resources such as employment and natural resources. The population increase will also contribute to environmental degradation which impacts on climate change, and other factors that cause environmental disasters such as pollution among others.

2.4. DEFORESTATION

Rain forests all over the world are still being cut down but does this make a difference? Should we care is some animals or plants in the ecosystem die? The concern should not be a few animals or plants but humans stand to lose more with the continued destruction of forests. Environmental degradation is a contribution of the human beings which comes back to bite us as by doing this we risk our own life quality and also affecting stability of the climate, threatening biodiversity, and also undermining the services provided to us by our ecosystem (Figure 2.5).

Figure 2.5: Deforestation.

Source: https://impakter.com/reforestation-the-new-norm/.

Despite the fact that environmental degradation is yet to reach a level of crisis in which most of the ecosystems collapse, it is still important to pay attention to existing environmental problems and try to mitigate some of these effects. Continued loss of ecosystems that are natural increases the vulnerability of human activities to ecological disasters in the future. Deforestation has significant impacts at local level when there is loss of ecological provided by forests especially tropical forests. This forest provides habitat to wild animals and some indigenous communities. They also provide valuable services to humans and the ecosystem by prevent soil erosion, controlling floods, filtration of water, protection of fisheries and pollination function. This functions or services are important particularly to the world's poor who rely on natural ecosystems for their livelihoods. Deforestation on the other hand decreases the renewal of resources such as timber, plants that are medicinal, food such as fruits, nuts, and game meat. Deforestation has also had long term impacts as it affects the climate globally and biodiversity. The greatest challenge is observing these changes since it is easy to believe what we see. Science and technology has made this possible. The changes are challenging to observe from local effects as they take place over a long time scale and thus can be challenging to measure (United Nations, 2010).

There are local and regional consequences to deforestation. The local effects occur almost immediately. With the loss of forest ecosystems, the provision of valuable services that are underappreciated by humans are stopped or reduced that are forests ensuring the regular flow of clean water, protection of the community from disasters such as floods and drought. The forests infiltrates rainfall brought about by tropical storms while anchoring soils and release water at regular intervals. Tropical forests help to prevent and moderate destructive and drought cycles that occur in the event forests are cleared.

When deforestation occurs forest cover is lost, runoff as a result of rainfall rapidly flows into streams and elevates river levels and subjects' downstream villages, cities, and agricultural fields to flooding especially during the rainy season. During the dry season, similarly the areas downstream are the most affected by drought that takes months which interrupts the navigation of rivers that wreaks havoc on crops disrupting industrial operations (Coppola, 2010).

Forests are mostly situated on steep slopes, montane, and watershed forests are important in ensuring water flow and inhibition of erosion, yet during the 1980s periods the montane forests suffered the highest deforestation rate of tropical forests a trend that has changed in the 2000s when upland forests recovered, while low land areas bore the brunt of deforestation, especially due to agricultural activities.

Forests also functions to add local humidity through transpiration and therefore adds to the local rainfall. For instance, 50–80% of moisture from the central and western Amazon remains in the ecosystems water cycle. In the water cycle, moisture is transpired and evaporated in to the atmosphere to form rain. When deforestation is done, the above process does not occur and there is a subsequent loss of rainfall leading to drought in the area. If the rain pattern changes for a few years the area then becomes arid, a scrub land and finally a desert over the course of time. River flow in this areas also decline, and small amounts of quality water reaches cities and agricultural land. The effects of change in vegetation forests to grasslands which do not favor rainfall consequently affect rainfall patterns. This change precipitation patterns have been observed globally for example Colombia which was one of the top countries with freshwater reserves is now falling at number twenty-four due to extensive deforestation, similarly deforestation around the Malaysian capital, combined with dry conditions created by El Niño triggered water rationing and in 1998, they had to import water for the first time as a result (Coppola, 2010).

There is also a serious concern that widespread deforestation could affect surrounding areas by increasing desiccation by the reduction of moisture stocks and the desiccation in extended regions. There is also a possibility of the desiccation extending to agricultural zones and other watersheds. As study done in Edinburgh suggests that in the next 50 years, a shift in the precipitation patterns which is induced by global warming and land conversion could lead to loss of the Amazon forest.

If a forest becomes dry or desiccated it, will then become prone to forest fires and spread to areas inhabited by humans which can be disastrous. Examples of these fires have occurred in different regions in 1997 and 1998 in relation to the dry conditions created by the El Niño. Millions of acres of land were burned in Indonesia, Brazil, Colombia, Central America, Florida, and multiple other places (Coppola, 2010).

According to the Indonesian National Board for Disaster Management (BNPB), the number of landslides increased to 73 from 29 in the period of 2001–2009. There were recorded from 18 to 216 flood cases and 198 cases of drought were recorded in 2008, which is the times the number recorded in 2003 which was 66 cases. The reduction of deforestation is important in the prevention of small scale disasters which include floods, landslides, and droughts.

2.4.1. The Climatic Role of Forests

The forests play a role in regulating the weather globally since they play an important role in precipitation and also in the exchange of atmospheric gases. Deforestation changes the reflectivity of the surface of the earth and this affects global weather by altering wind and ocean currents. It also changes the distribution of rainfall. The more the destruction of forests the further there will be change in global weather patterns hence becoming more unstable and extreme. The loss of forest vegetation cover implies that there will be less absorption of heat and this translates to a reduction in the amount of moisture taken up in the atmosphere. Precipitation is also affected when fires clearing forests create pollution and this leads to the release of tiny particles referred to aerosols to the atmosphere. Depending on their size, shape, and color aerosols can either have a cooling or heating effect on the air. The high concentrations of bio-mass burning aerosols have a direct impact on climate by increasing the formation of clouds but consequently decreasing the amount of rainfall according to the research done by NASA. Smoke surrounded areas have cloud droplets form around the aerosol particles, but do not grow large enough to fall as rain. Forest fires also have the effect of reducing rainfall further by leaving areas burned making them more prone to dryness and fires in the future (Dufresne, 2019).

This explains why areas of deforestation have a decline in the amount of deforestation. This affects other parts of the world by reducing deforestation for example deforestation of the Amazon in South America by influencing rainfall from Mexico to Texas and the Gulf of Mexico and also the loss of forests in central Africa affects rainfall patterns in upper and lower U.S. Midwest. Deforestation in Southeast Asia had an impact on precipitation in China and Balkan Peninsula. Two Russians scientists in 2007 proposed the pump theory that explains the role of forests in the generation of forests on masses of land. It proposes that condensation in forests and not precipitation differences drive winds to bring precipitation over land and this was contested (Dufresne, 2019).

2.4.2. Atmospheric Role of Forests

Forests play an important role of absorbing carbon in the atmosphere as they use it in the process of photosynthesis. The degradation and clearing of forests has an effect which is opposite to the benefit. Large amounts of carbon released remain in the atmosphere together with other greenhouse gases such as nitrous oxide, methane, and others. This buildup of carbon is referred to as the greenhouse effect. This accumulation alters the world radiative balance leading more absorption of the heat from the sun inside the earth's atmosphere and the result is global warming. Greenhouse gases for instance carbon IV oxide have a transparency towards short wave solar radiation. This radiation upon reaching the earth surface heats it and radiates it as long wave radiation. The heat is therefore trapped in the atmosphere. With the buildup of greenhouse gases, the opacity is increased and there is more heat trapped within the atmosphere. CO_2 makes the largest contribution to the greenhouse effect and is a result of fossil fuel combustion before industrialization the levels of carbon dioxide were at 280 parts per million but have risen to 400 parts per million which has an increase in temperature as a result (Dufresne, 2019).

Despite this, there are still a lot of factors unknown about its impact on climate change and has been a constant subject of debate. Some of the effects include an increase in temperature at the north and South Pole, breaking of ice sheets in the Antarctic, receding of glaciers in the world, the longest El Nino among, heat waves, an increase in epidemics which is attributed to climate change globally among other environmental disasters. The Intergovernmental Panel on Climate Change (IPCC) projects that the rise in carbon IV oxide levels predicts that the levels could reach between 450 and 550 parts per million in the years 2050. Resulting in even higher temperatures globally and contributing to the possible rise in sea levels along with a range of impacts that range from increased storm and hurricane intensity, melting of polar ice caps, changes in ocean currents, a rise of global sea levels which could inundate low elevation cities like Cairo, Venice, logos, New Orleans and Amsterdam. This effect can cause problems for nations that are low-lying, bleaching of coral reefs and mortality of coral reef ecosystem which act as a barrier of protection against tsunamis. It also leads to the migration of species, extinction of species especially species within cold climates. There is also increased danger from pollutants to humans such as ozone, the spread of tropical diseases in cooler climates and other pathogens; and shortages of water (Coppola, 2010).

2.4.3. Solutions to Deforestation

- Rehabilitation of areas that are close to, within or already existing forests
- Tree planting through afforestation and reforestation-people should be encouraged to plant trees in their areas where forests were cut down
- Raising awareness on the consequences of deforestation as most people put their focus on economic gains and do not put into

account the long-term environmental damage and this attitude has encouraged the illegal logging of trees.

- Youth involvement-the youth are tomorrows future and instilling values such as forest preservation, afforestation, and reforestation will help sustain forests even for future generations. This can also be done by involving them in tree planting activities especially within their community.
- Individual tree planting-people should be encouraged to plant trees in their vicinity especially in areas they encroached
- Development of rules and regulations that prevents deforestation and encourages afforestation they include the banning of tree cutting to curb the depletion of forest cover and this solution is feasible. Organizations and governments work together or should work to enforce these rules and legislations to preserve forests.
- Land use planning-as more and more people move to urban areas, they encroach forest land for settlement and agricultural activities. Proper land use planning techniques provide fast and feasible solutions to deforestation. The techniques are centered on techniques like agriculture that is urban and lessens the urban and suburban sprawl and this can reduce deforestation activities.

2.5. RISING SEA LEVELS

Rise in sea levels and the project rise in sea levels from expansion of ocean water and melting of the polar ice caps is dependent on global warming estimates. There is an estimated rise in oceans about 10 inches to 20 inches if there is continuous rise of greenhouse gases. A rise of such levels may not seem to be significant but it may affect human and natural ecosystems. Any increase in sea levels may be magnified in the event of tides, storm surges, and even hurricanes and this could lead to an impact that is devastating as shown by hurricane Katrina in 2005. Nations on islands face extinction for example Maldives and south pacific which is scattered. The sea offers important resources to humans whom they use to sustain their livelihoods but at the same time, it can pose a threat to their lives. A rise in sea levels can affect this by causing flooding and causing a potential disruption of sewage and transit system and along with other agricultural plots. Other than causing disasters the change in sea levels affect coastal ecosystems and destroy low lying forests which also are beneficial to man in addition to acting as a habitat to biological diversity. Although sea levels have risen in the past, currently there is space lacking for aquatic species to move to in times of flooding as these areas which were extensions of their environment are inhabited by humans. It takes a short duration and this is dependent on the tides (Paul, 2011).

There are three types of floods and they include flash floods, river floods, and tidal floods. Flash floods are of small duration but take a small period. River floods occur as a result of heavy rainfall aver large area, snow melting or both. Tidal floods are overflows in coastal regions which are caused by hurricanes, high tides, storm surges, high sea waves, tsunamis, and a combination of both.

2.5.1. Effects of Floods

The major effect of the occurrence of floods there is often destruction of complete destruction of infrastructure in an area. Even a small flood can cut major highways, cause overflows of sewage systems and this can lead to the contamination of clean water sources. Other direct damages also include injury to people, loss of lives and damage to structures that are man-made as a result of swift currents, debris, and sedimentation. Loss of soil and vegetation can be caused as a result of erosion of sedimentation and deposition on landscapes.

Short term pollution is an indirect impact or effect of flooding, disruption of food supplies, disease spread, and the displacement of people.

2.5.2. Control of Flooding

There are different measures that can be employed and applied to prevent and control floods. The measures include the construction of levees and flood walls to function as barriers against water that is high. Building of reservoirs to hold the excess water from the flood until it is a safe time to release the water, expansion of channels to facilitate the movement of water quickly of the land and the diversion of the channels of flood water around areas that require protection from floods. Engineering controls and flood plain regulations should be adhered to in order to control floods. People should avoid building houses on flood plains unless it is elevated. Elevation of electric poles in areas susceptible to floods.

2.6. SOIL EROSION

Soil erosion is a process that involves the top soil wearing away. Soil particles lead to the loosening of soil particles and is washed away to oceans by rivers or streams. Soil erosion is a process that is natural but is exacerbated by human activities such as deforestation and agriculture. Soil erosion agents wear away top soil they include force that are natural such as water and wind which contribute to annual loss of soil. This is a continuous process that can occur at an alarming and an unnoticed rate and has a contribution towards the loss of large amounts of top soil. Loss of trees by deforestation may cause widespread erosion since it functions to anchor the soil with the use of its roots. Another effect of soil erosion is the loss of soil fertility which is lost in the event of flooding or heavy rains.

2.6.1. Causes of Soil Erosion

Soil erosion goes on globally but some are more vulnerable than others due to different causal factors such as human activities or natural causes. Some of the causes include:

- Rainfall and flooding: greater events in soil erosion are associated with rainstorms. It produces different types of soil erosion such as rill erosion, gully erosion, sheet erosion, and splash erosion. The types of erosion are result of the impacts caused by rain on the surface of soil that are washed away by runoff from storm water. This leads to loss of significant amounts of soil. Rapid storm water movement, flash floods and flooding occur as a result due to excessive run off thus causing extreme local erosion by bed rock plucking and washing away of loosened soil particles.
- Rivers and streams-flowing rivers and streams cause valley erosion. They eventually lead to valley erosion. When the flowing water eats away the soils and provides v-shaped erosive activity. With the rivers of soil deposits as a result of sedimentation, it also levels up with the surface and the water ways begins to wash away the soils at the banks. This activity of erosion and is described as lateral erosion. It is evident during heavy rainfall and rapid river channel movement.
- High winds-they contribute to soil erosion, especially during periods of dry weather or occur in arid and semiarid regions. The winds with its natural force picks up loose soil particles and carries them away to different areas leaving the land bare. Soil erosion

is severe during the drought periods in the arid and semiarid land (ASAL) regions and degradation and desertification are the result of wind erosion.

- Overgrazing, overstocking, and over cultivation of arable land-human encroachment for the purposes of cultivation and livestock keeping has contributed largely to soil erosion and top soil loss. The human activities have led to reduced cover on the ground and break down of the soil particles, allowing erosion and the acceleration of the effects of wind and rain.
- Deforestation, reduced vegetation cover and urbanization-vegetation and urbanization is a way to destroy vegetation cover. They include agricultural practices which include burning and the clearance of vegetation which function to reduce the cover of vegetation. This loss of vegetation cover leads to an increase in the rate of soil erosion. Erosive effects can be reduced by trees and vegetation that can be caused by rainfall and flooding. Human actions such as deforestation and urbanization lead to the continued loss of soil.
- Mass movements and soil structure-in the soil erosion process, the outward and downward movement of sediments and rocks on slope surfaces as a result of gravitational pull. This mass movements aids soil particle breakdown making it vulnerable to water and soil erosion. The erosive ability by wind and rainfall is determined by soil structure and composition. For example, clay soils have the tendency to be resistant to soil erosion in comparison to sandy and loose silt soils. Moisture content in soil and organic content are some of the characteristics are some of the aspects that will determine soil erosive ability of wind or rainfall.

2.6.2. The Effects of Soil Erosion

The effects of soil erosion are centered on the reduction of agricultural productivity as well as soil quality. These include the blockage of water ways and this may affect the quality of water. This means most environmental problems are as a result of soil erosion. They include the:

- Loss of arable land-agricultural land can be affected by soil erosion. This carries away the top soil that is fertile and also the components that support the soils organic matter therefore this result threatens the productivity of arable land as they are continually degraded. The characteristics of soil that support agriculture are degraded by soil erosion leading to collapse of soil ecology and starvation of masses.

- Water pollution and clogging waterways-eroded soil from arable land carries with it pesticides, heavy metals, and fertilizers which are washed into streams and major water ways. This leads to the pollution of water and cause damage to habitats that are marine or freshwaters. The accumulation of sediments leads to the clogging of water ways leading to raised water levels which leads to flooding.
- Sedimentation and threat to aquatic systems-other than the pollution of water systems, sedimentation can be of catastrophic to the survival of aquatic life forms. Silt affect fish breeding grounds and lessens the food supply since the presence of silt reduces algal life biodiversity and is of benefit of aquatic life. These sediments can also enter gills of fish affecting their respiratory functions.
- Air pollution-air pollution is caused by wind erosion which picks up dust particles and spreads them into the air. These dust particles can bring harm or be toxic such as petroleum and pesticides that are health hazards in the event that they are inhaled or ingested by humans. As the wind moves air pollution is the resulting effects especially in dust plumes from deserts or dry areas.
- Destruction of infrastructure-infrastructural projects such as dams, drainages, and embankments are affected by soil erosion. Soil accumulation reduces the efficiency and lifetime of such projects. The silt also supports plant life that can cause cracks and the weakening of structures.
- Desertification-desertification is as a result of soil erosion. Desserts are the result of transformation from land that is habitable. This transformation is made worse by destructive use of land and deforestation that leaves the soil bare. This leads to biodiversity loss, the alteration of ecosystems, degradation of land and also huge economic losses.

Soil erosion is a man-made disaster; this is because whatever human beings build alters the natural regime in ways that have environmental consequences that are disastrous. It is only a problem in the event that it is accelerated or in excess by alteration of the earth's landscape, alteration of drainage patterns, changes of the contours of the earth, building roads without the regard of storm runoff, and paving everything in sight resulting to peak runoffs that are high in volume and velocity and faster in arrival. Humans also tend to construct houses in areas that are prone to erosion which is deemed to be of high value. Impacts of soil erosion to both humans and the environment include:

- Safety risks are caused by excessive erosion and this can undermine constructed infrastructure such as roads, bridges, and building foundations.
- The destabilization of hillsides by soil erosion triggers landslides that would have otherwise not occurred if it was not for the impact of human activities on the location, rate, and the runoff volume.
- The buildup of silt and mud in water reservoirs is as a result of excessive erosion which are important in the control of floods, energy generation and the supply of water.

2.6.3. Measures to Mitigate or Prevent Soil Erosion

- **Planting of Vegetation Cover:** People are encouraged to cover and soil binding. Plants actively prevent erosion by wind and water by providing cover to soil by binding it with its roots. Small trees, herbs, grass, and wild flowers are some of the plant choices that can be used. Other plants that are excellent in soil erosion prevention are creepers because they spread to the ground instead of growing straight.
- **Application of Mulches:** Another soil erosion control strategy is by mulching of soil. Putting dead leaves and shredded word on the soil is how mulching is done. The process of mulching provides cover for trees and plants against weather that is extreme. Mulching allows the slow reach of water and thus this reduces the rainfall impact or heavy watering. It functions to prevent turning acid from soil and this leads to the suppression of growing weeds. As time goes these mulches which are made from organic matter break down and lead to an increase in soil structure and fertility.
- **Matting the Soil:** It is a process of soil covering biodegradable materials which includes straw, coconut fiber, and wood chips. These materials are kept together by netting that is biodegradable made of coco coir. This is the equivalent of covering the soil with a carpet which in this case is biodegradable. Spikes that are

biodegradable function to hold the matting material. The material decays and is converted into fertilizer. The use of plastic is not acceptable as it is not biodegradable.

- **Contour Farming and Terracing:** To reduce soil loss from surface erosion, the practice of tilling land sloped along lines of consistent elevation functions to conserve rain water. Contour farming act as reservoirs that catch and retain water. This action permits increased water filtration and a more uniform distribution of water.
- **Creation of Windbreakers:** These entails the planting of trees around gardens or within farmlands a practice commonly referred to as agroforestry. It prevents the blowing away of soil particles by wind and water. The trees prevent and help in the protection of crops from damage and this reduces the loss of soil by surface erosion that can be caused by wind and this improves the health of plants and the quality and quantity of agricultural produce. Protection of crops from wind is of significance in the retention of moisture by plants.
- **Crop Rotation:** It is important for keeping arable land healthy. This allows the buildup of organic matter providing fertility for future agricultural produce.
- **Water Control:** In most areas soil erosion is mainly caused by water either naturally or manmade. Run off pipes and chutes that are specialized can help in directing these sources of water away from areas that are susceptible to soil erosion, by helping to stave off erosion that is in excess. Employing the filters in particular areas rather than driving water to natural bodies of water works to reduce pollution.
- **Increase Knowledge by Creating Awareness:** A major prevention strategy in combating soil erosion is creation of awareness about the problem and educating more and more people who work with the land. It starts by explaining why it is a concern and also providing solutions to help reduce it. This means the provision of outreach services especially to farmers in susceptible areas including ways that can help farmers in areas of susceptibility to erosion in ways that can provide protection to crops from bad weather, or other ways that assisting in making

sure that soil remains compact but does not affect agricultural activities all the same.

- **Building Retaining Walls:** A wall retaining structure is a structure that functions to keep, soil, rocks, and water in place so that it is not eroded by rain.

2.7. RAPID URBANIZATION

The world is seen to be experiencing a shift in peoples' settlement as a large part of the world's population now lives in urban areas after decades of migration from rural areas. It is estimated that by 2050 thee figure will surge to 75%and developing countries account for a large portion of the migration. As much as urbanization creates new opportunities for growth and strengthening of the economy it is also responsible for putting a strain on both natural and monetary resources. This makes urban centers more susceptible to disasters both manmade and natural.

A large majority of the people who migrate to these urban areas are poor. Currently more than 800 million people globally live in slums and this includes more than 60% of city dwellers in sub-Saharan Africa. Rapid urbanization has the potential to increase national wealth according to UN habitat. Informal mechanisms of land acquisition, building, and infrastructure have led to urban growth in many low and middle-income countries. In urban areas, both the informal economy and settlement are commonly associated with increased exposure and low pay which risk exposure to environment hazards. Therefore, low, and middle income countries have a disproportionate concentration in mortality and the risk to environmental disasters (Smith, 2003).

With the growth cities, there is increased risk of natural disasters. The disasters range from fires, landslides, floods to tidal waves. The rise of such disasters is alarming. The continuous movements of people to less resilient areas like coastal regions, flood plains, and zones that are prone to earth quakes have been cited as one of the factors considered to be responsible for disasters that are natural. Environmental degradation is another major reason to natural calamities and a result of the urbanization process. Expansion that is unplanned in disaster prone areas in another reason for the increased number of disasters. The center of research on the epidemiology of disasters has made the observations on the number of people affected by disasters.

The rise in numbers of people affected can be attributed to the growing population of the world and evolving reporting standards. With the changes, there is also the growth of global awareness and the unwillingness to tolerate the extreme suffering as a result of natural disasters.

According to some reports, the reports of major floods have increased off less than fifty to just below two hundred per year. The incidences of tropical storms have increased from ten to fifty to below a hundred in a year.

The growth of cities threatens to make them unsustainable. In many developing countries, city authorities have difficulty in the provision of basic infrastructure and services. The result leads to a large percentage of people in some of the largest cities to live in densely populated areas in developing countries (Smith, 2003).

With the growing population and the increase in demand for land in cities, has led to settlement on unsuitable terrain such as floodplains, unstable slopes, reclaimed land that can be prone to natural hazards. Urban development increases the risk of floods as the result of the disruption of natural drainage channels. Cities that are fast growing contain an increase in the number of buildings that are constructed poorly and also these buildings are badly maintained and can lead to deaths that are unnecessary.

2.7.1. Effects of Rapid Urbanization

Urban development can alter the ecosystem and the environment. In urban areas the expansion of paved and impermeable areas, prevent the absorption of rainfall which would have been infiltrated to the ground. Therefore, increasing the probability to flood hazards especially in areas that are low-lying. Cities that are not well planned and managed create risks that can be negative developments to the gains made. The lack of inadequate infrastructure and services, housing that is unsafe and poor health services can trigger natural hazards that can lead to disasters. For example, poor solid waste management can lead to storm water blockage and sewage networks that can lead to water logging or flooding. Water scarcity and contamination is a result of destruction and damage to infrastructure. When there is lack of proper housing and water provision, sanitation, healthcare, and education affects the recovery of urban residents.

Urban development has also led to the destruction of natural barriers, including swamps, wetlands, along rivers and mangroves where there is fertile agricultural land and transportation. Major cities around the rim of the Pacific Ocean especially those close to active fault lines and have a significant risk to earth quakes. Soft soils that mostly accumulate along river banks and delta can cause amplification to ground motion experienced during a temblor and may lead to damage by liquefaction. Areas that are reclaimed for development can also lead to liquefaction (United Nations, 2010).

The increase movement and settlement of people in a majority of coastal regions have increased over the years. With the population, growth around these regions there is also the aspect of property development. Low lying regions are occupied by communities in coastal areas. These low lying regions have direct sea or ocean access exposing them to damages by wind or water which is also magnified by rising sea levels and subsidence of land. The rising sea levels and severe weather events in the coastal regions are associated with an increased risk of flooding.

In urbanized areas, there are a number of homes that are developed within the urban and wild interface as a result of increased population pressure. This is where urban expansion meets vegetation that is natural and in such locations, it can be hard to prevent and suppress fires. The risk of fires can be exacerbated by different factors including fire extinction as fast as possible. Wildfires lead to huge losses of property, housing, and lives. Despites the cases of wildfires not being a common occurrence in many countries there is a steady occurrence in countries where they occur such as the United States. This occurrence is attributed to the growing value of properties exposed (United Nations, 2010).

2.7.2. Proposed Solutions to Urbanization

- **Building of Sustainable Environmentally Friendly Cities:** Legislators should create laws that plan and provide environmentally urban regions and smart growth techniques. This is in consideration that people should reside in areas that are safe and free from pollution. The objective of this is to build cities that are sustainable and embraces environmental conditions that are improved and habitats that are safe. The government should also encourage the sustainable use of resources and support an economy based on sustainable environment such as the investment in infrastructure that is green, industries that are sustainable, environmental campaigns on recycling, management of pollution, energy that is renewable, green public transportation and recycling of water.

- **Provision of Essential Services:** Governments and urban stakeholders must ensure people living in urban areas are provided with social services that are essential. The services include health and sanitation, education, technology, clean water, electricity, and food. Another objective is the creation of job opportunities and wealth creation activities so that the population can earn a living to pay for the maintenance of services. The government can also provide subsidies to lower the health care cost, education, energy, education, public transportation, systems of communication and technology.
- **Creation of More Jobs:** This will function to reduce the effects of rapid urbanization and also it will help in the conservation of ecosystems that are natural. The promotion of tourism and the sustainable use of natural resources can lead to the creation for the growing urban population. Provision of subsidies and grants may be provided to foreign investors in the development of environmentally friendly projects that also encourages the creation of jobs.
- **Population Control:** Key stake holders in urban areas should do campaigns and provide health services especially family planning to reduce the high population growth.

2.8. POLLUTION

Pollution is the process of introducing contaminants in to the environment that can have adverse environmental and health consequences. Pollutants take the form of chemicals, energy, noise, heat, and light. 2015 pollution so the death of 9 million people and this is fifteen times high that deaths resulting from deaths caused by war or violence or three times the number caused by HIV/AIDS. Pollution is one of the greatest anthropogenic factors. The major forms of pollution include air, land, water, noise, radioactive, thermal, and light pollution (United Nations, 2010).

Industrialization has seen the birth of different forms of environmental pollution. With industrialization there is continuous accumulation of waste generated that has led to environmental pollution. This is a problem occurring at a global scale. The largest amount of waste generated by environmental disasters such as volcanic activity, earthquakes, forest fires, tornadoes, tsunamis, flooding, tsunamis among others. The waste is generated in large amounts for a very short period of time it includes both managing the effects at environmental and economic measures. If the effects are not managed almost immediately, it can result in environmental disasters and become a public health concern.

Environmental pollution of the air and water causes a lot of deaths annually at a global level. It kills many people in comparison to disasters, hunger, and natural disasters. In 2015 statistics, for global mortality rates it was reported that one in six premature deaths were caused by diseases that were as a result of exposure to toxic substances released by industrial activities. This was according to the lancet journal. The financial implication of treatment and welfare is equally massive. Despite pollution being a huge problem, it is yet to receive the attention it deserves like other factors causing death. In 2015, it was linked to 9 million deaths. Polluted air is as a result of fires, natural disasters and industrial activities were linked to 6.5 million deaths. The next and largest cause of death was water pollution which led to gastrointestinal diseases and disorders, parasitic infections which killed 1.8 million people (United Nations, 2010).

When pollution occurs, at once it can lead to disasters that lead to the loss of lives or have adverse health effects. There are several international catastrophes that have occurred as a result of accidents that led to the pollution of the environment. An example is the Exxon Valdez oil spill. Or the Amoco Cadiz oil tanker of Brittany and the Bhopal disaster in 1984 have demonstrated the universality of such events. The atmosphere and water has a borderless nature and the implication of pollution can be felt at a planetary level with the global warming issue. Another global issue is waste disposal which contaminates oceans for example the great pacific garbage patch that is comprises of plastics, chemical sludge and debris which has been collected into a large area of the Pacific Ocean by the north pacific gyre (United Nations, 2010).

2.8.1. Effects of Pollution

Human health-environmental air pollutants can lead to the death of biological organisms. The pollution of the ozone layer causes respiratory, cardiovascular disease, chest pain, throat inflammation, and congestion. Water pollution can approximately lead to 14000 deaths on a daily basis as a result of the contamination of clean water sources for example as a result of the release of untreated sewage especially in developing countries. For example, in India in 2013 over 10 million people fell ill as a result of waterborne illnesses. In India, air pollution caused an average of 1,535 deaths. In china in the year 2010, the estimation of premature deaths was 1.2 million as a result of air pollution. Environmental contamination with mercury has been linked to neurological defects, developmental deficits. Older people exposed to air pollution and those with heart or lung disorders suffer an additional risk. Lead and other heavy metals cause neurological problems to old people. Radioactive and chemical substances can lead to cancer and at the same time birth defects. Oil spills also cause skin irritations upon exposure (Weil, 2004).

Environmental effects include bio magnification where toxic pollutants exceed trophic levels. Emissions of carbon IV oxide lead to acidification of the ocean. Emissions of greenhouse gases results in global warming and this affects the climate and the ecosystem in different ways. It contributes to the loss of biodiversity. Nitrogen oxides change the species composition within the ecosystem. Smog and haze reduce the amount of sunlight by plants to facilitate photosynthesis and also leads production of ozone in the troposphere and this causes damage to plants. Pollution together with soil erosion can lead to soil infertility making them unsuitable for plants, this goes further to affect organisms within the food web. Pollutants such as sulfur dioxide and nitrogen oxide result to acidic rain which lowers soil PH value. Organic pollution depletes oxygen in water sources which reduces aquatic biodiversity.

2.8.2. Control of Pollution

Environmental management is a step in pollution control. It entails controlling emissions and effluents into the air, soil, and water. Without management, waste products accumulate and lead to environmental degradation. Preventing pollution and minimizing wastes are more desirable measures in the control of pollution. The practices include recycling, reusing, minimization of waste, mitigation, prevention, and composting.

2.9. OZONE LAYER DEPLETION

The ozone layer can be defined as a region within the Earth's stratosphere that absorbs radiation from ultraviolet (UV) radiation. In comparison to other parts of the atmosphere, the stratosphere contains the highest concentration of ozone. It is location in the lower portion of the stratosphere, and it is around 9 to 21 miles (about 15–30 kilometers) above Earth's surface. Ozone layer depletion is defined as the thinning of the earth's ozone layer. The process occurs gradually in the upper atmosphere. This depletion is caused by the release of toxic substances that have a bromine or chlorine component from both industrial and human activities. Ozone depletion can bring serious problems to the environmental system as it causes adverse effects (Coppola, 2010).

What is the importance of the ozone layer? The ozone layer prevents exposure to UV radiation that can be harmful to human health as it leads to serious diseases and cancer. It does not only bring harm to humans but also plants and animals. The radiation can affect animals by decreasing their fertility rate and also a reduction in the growth of plants exposed to ultra violet rays. The ozone layer not only protects biological creatures but also the environment.

- **The Montreal Protocol:** Montreal Protocol is an international treaty formally enacted in Montreal on September 16th, 1987, with the goal of regulating production and usage of chemicals that cause destruction of Earth's ozone level. The global treaty was developed to safeguard the ozone layer through removing the production of various substances thought to be liable for ozone depletion.

It proposes precautionary measures to regulate equitably total universal emissions of elements that deplete it, having with the ultimate goal of their total elimination on the context of improvements in scientific knowledge, while taking into consideration technical and economic factors, and having in mind the developmental requirements of developing countries.

2.9.1. Causes of Ozone Layer Depletion

There are several causes of ozone layer depletion. They are divided into natural and manmade causes. They include sunspots and can be defined as temporary phenomena on the suns photosphere and as regions of reduced surface temperature caused by magnetic field flux and leads to the inhibition of convection. Sunspots bring along an increase in UVB-radiation which alters concentrations of the ozone in the atmosphere. Stratospheric winds naturally lead to ozone layer depletion. It is a natural cause of ozone layer depletion. They lead to an increase in nitrogen oxides within the atmosphere and this further leads to further depletion of ozone layer. Another cause is volcanic eruptions. The role they play is indirect in the depletion of the ozone. Volcanic emissions produce large amounts of aerosols and particles when they erupt. The particles that are emitted by volcanic activity create a surface on which chemical reactions can take place and this can bring harm to the ozone layer. Chlorofluorocarbons have often been used globally for purposes of refrigeration in the past. Their use has declined with the discovery of its harm to the ozone layer and its contribution to global warming. Halons are also known for their contribution to ozone layer depletion since it contains bromine and has the potential to destroy ozone within the stratosphere. In comparison to chlorofluorocarbons, halons can pose a threat ten times harmful to the ozone layer. Previously halons were used as fire extinguishers (Coppola, 2010).

Although they are more harmful, chlorofluorocarbons are more commonly used. Thus, the impact those chlorofluorocarbons to the ozone layer are greater than that of halons. Carbon tetrachloride contributes to the depletion of ozone layer since it contains chlorine that leads to depletion of the ozone layer. It is a substance that is also used for refrigeration purposes among other functions. Its use has also declined over the course of time. Methyl chloroform which is used in industrial process and leads to ozone layer depletion. The UV rays break up the substance into chlorine which then goes further to bring harm to the ozone layer. Methyl bromide is also another substance that leads to ozone layer depletion and is used as a fumigant for pest control. The substance is classified as a class I ozone depleting substance. Other substances include hydro fluorocarbons. The ozone layer is also depleted by human activities examples include the use of fossil fuels that produces carbon emissions which also lead to the depletion of the ozone layer (Babar, 2007).

2.9.2. Effects of Ozone Layer Depletion

Increased ultra violet radiation is one of the effects of ozone layer depletion which as has severe effects to all life forms. The health effects include cancer with threatens human life. It also causes melanomas which is a form of skin cancer as a result of UVB exposure. The effects to human health include sunburns, cataracts, aging, and also the weakening of the immune system leading to other diseases. A depleted ozone layer affects animals, such as marine life. It can also cause cancers to animals and can have adverse effects because unlike humans they do not employ protective measures from the sun or rather ultra violet rays. It also affects plants especially their growth. Plants inhabiting the earth are used to certain conditions within the environment. In the event they are exposed to different environmental conditions, their reaction may be sensitive for example an increase in the exposure to UVB radiation. It also leads to a reduction of crop yields by alteration of plants DNAs. Ozone layer depletion has several adverse effects to the environment. It is difficult to provide an estimate the magnitude of the effects that ozone layer depletion has on the environment as many reactions are included in the process. It is estimated that without the presence of the ozone layer, collapsing of the food chain may occur and therefore measures should be put in place in ozone layer preservation. It also has an economic impact as it causes a down side of the economy on a global scale.

There are two regions where the ozone layer has been depleted:

- In the mid-latitude-for example over Australia, the ozone layer has been thinned leading to an increase in the amount of UV radiation reaching the earth; and
- In the atmospheric regions over the Antarctica, the ozone layer is significantly thinned especially during the spring period and has led to the formation of ozone holes. One of the largest ozone holes makes it appearance yearly between September and November of the Antarctica.

Research done by Professor Kang of the Ulsan Institute has shown that depletion of the ozone layer at the South Pole has affected the extreme daily precipitation in the Australian summer, for December, January, and February (Kapur, 2010). The ozone hole over Antarctic affects atmospheric circulation in the Southern hemisphere. Ozone layer depletion is associated with extreme rain in the austral summer and it would be used in forecasting of heavy rains and natural disasters in the future.

In the Australian state dangerous floods have occurred over the last 150 years. These disasters have led to the loss of life, destruction of infrastructure.951 have been killed by floods between 1852 and 2011, 1326 people were injured and the damage incurred losses of up to 4.76 million dollars. Even though natural disasters can be predicted, the disasters cannot be stopped. The measures to be employed include preparation and the reduction of damage (Babar, 2007).

It is hard to determine changes of precipitation as there is limited availability of data to support this. Recent changes in climate at the southern hemisphere that is as result of ozone holes which can be observed with the recent change in climatic patterns.

2.9.3. Solutions to Ozone Layer Depletion

- **Avoid the Use of Pesticides:** Pests and weeds especially on agricultural land are destroyed by chemicals referred to as pesticides. These chemicals contribute to the depletion of the ozone layer. The only way to achieve this is by the use of natural methods to kill pests and weeds is the employing of natural methods or the use of chemicals that are friendly to the environment. Manually weeding of farms can also be used.
- **The Use of Renewable Energy:** To avoid the continued depletion of the ozone layer, it is important that people adopt the use of renewable energy sources and avoid the use of fossil fuels, nuclear energy and coal which are important in the generation of electricity. With emission of chlorofluorocarbons in the combustion of coal, the practice is rather discouraging and the use of fossil fuels encouraged but still limiting its use.
- **Putting a Stop to Deforestation:** Deforestation is a measure that indirectly mitigates ozone depletion since cutting down trees contributes to global warming. Global warming contributes to ozone layer depletion by increasing the amount of atmospheric water vapor hence blocking the production of ozone. When we burn wood, the substances that result from the process contribute to the depletion of the ozone layer
- **Recycling and Reusing Products:** It is advisable to either reuse or recycle products as much as we can. In the event that we no longer use them, people should also try to recycle these products in a proper manner. By doing this we can reduce the demand of natural resources. The adverse effects that result from the extraction of natural resources from the environment can be avoided. When there is reduced demand for energy the resultant effect is a decrease in carbon emissions and thus the depletion of the ozone layer is lessened.
- **Avoid the Excessive Use of Fertilizers:** Many fertilizers contain nitrogen and other substances that are harmful. The excessive use of these chemicals can cause harm to the ozone layer. It is important to use fertilizers in a responsible way by not overusing them. This is a measure that can be applied through the use of government regulations and also the provision of education to farmers on the effect of excessive use of fertilizers.

- **The Reduction of Chlorofluorocarbons:** As seen above chlorofluorocarbons are the major contributors to ozone layer distribution. These chemicals are mostly used in industries for production. This includes refrigerators, air conditioners, and aerosols. If the chlorofluorocarbons used can be replaced with other substances then these emissions can be reduced hence protecting the ozone layer.
- **Reduction in the Use of Nitrous Oxide:** Like CFCs, they are produced as a result of industrial processes. There are adverse effects produced that harm the ozone layer. The solution will be to reduce the amounts used or replace the chemicals.
- **Government Regulations:** Institution, the government, and non-governmental organizations have the responsibilities to implement measures that protect the ozone layer from being depleted. The responsibility of sustaining the ozone layer should be a collective effort and if the matter is not prioritized then there will future consequences that have an adverse effect to human health, plants, and animals. The government should ban ozone depleting substances and replace them with substances that are less harmful. The government can also impose high taxes on such substances to discourage their purchase and this financial incentive will lead to the purchase of substances that are less harmful to the ozone layer.
- **Creation of Awareness and Provision of Education:** To fight ozone layer depletion we can make a choice of educating individual. This begins with the creation of awareness on the harm brought to the ozone layer by the use of certain substances and also the consequences of the depletion of the ozone layer. Education can be done at local and industrial level. People can also be educated at a young age in schools and also allowed to participate in forum that involves the discussions about environmental issues. This knowledge acquired can then be passed on to generations to come and thus is a continuous process. Education will be successful if people adopt the policies and legislations that protect the ozone layers. This will also profit industries that adopt the environmentally friendly measures as customers will also be driven towards purchasing products that conform to environmental standards. This consequently contributes to a positive in relation to the environment.

2.10. TECTONIC PLATE MOVEMENT

Natural Disasters are distinct from environmental disasters. While environmental disasters are man-made, natural disasters such as those caused by tectonic plate movement can cause environmental damage but they are quite distinct as to what the root cause is.

Plate tectonics are forces involved in shaping the earth. The earth is made up of plates and not a single mass. These plates move slowly over the course of time and create changes to the surface of the other and its structures. These plates can move abruptly and when this happens, the earth's surface is subjected to natural disasters. The disasters include earthquakes, volcanoes, and tsunamis.

2.10.1. Earthquakes

An earthquake is the sudden shift of the earth at fault lines where at the crust displacement occurs. The sudden movement of tectonic plates along fault lines can lead to the occurrence of a lot of earthquakes. This movement sometimes does not occur smoothly. Then plates do not interact smoothly and as a result of friction, they may catch each other. As the plates move this may cause the buildup of energy along the fault lines. The energy is released and results in an earthquake. In California, there exists the famous San Andreas Fault and it marks the location where the North American plate and the pacific plate slide past each other. They move at a rate of approximately 6 cm yearly and results into tiny earth quakes and the event of a major earth quake occasionally. The size of the earthquake is measured by a seismograph and the damage expected can measure.

2.10.1.1. Effects of Earthquakes

The permanent effect of the earthquake is the destruction it causes in relation to the ground. This then causes damage to buildings, bridges, dams, tunnels, and other infrastructure. Long lower frequency shock waves can cause more damage in comparison to short sharp waves of higher frequency as the longer waves lead to liquefaction of the ground and therefore unable to provide support to the structure. On silt or sandy soils, liquefaction occurs when it has high water content or when groundwater is closer to the surface. secondary effects that are as a result of earthquakes include broken gas and water lines, fires as a result of damaged gas, release of chemical and radioactive waste, landslides, tsunamis, and floods. Landslides can occur when rocks fall out place and this occurs on steep slopes. Other effects include electrocution, drowning, and the occurrence of respiratory diseases and are as a result of exposure to dust or asbestos and can be a health hazard

2.10.2. Eruption of Volcanoes

A volcano is a physical structure and is a vent, fissure, or chimney found within the earth crust and through it, molten rock rises to the mountain surface. For most active volcanoes, they exist at the margin of tectonic plates. In nature, Volcanoes exist along plate boundaries. When there is plate movement over another plate, the energy produced and the friction causes melting that pushes magma upwards. The pressure beneath the molten rock and without another outlet to release the pressure and it explodes as a volcano. An eruption can also occur if magma oozes when there is pulling apart of plates and the space left behind is filled with the magma. This type of eruption is dependent on the molten rock beneath the surface. The type of eruption has occurred in Washington in 1980 on the Helen mountain other volcanic eruption occur slowly especially those that are in Hawaii (Figure 2.6).

Figure 2.6: Kilauea volcanic eruption in Hawaii.

Source: https://www.pinterest.co.uk/pin/839921399225413504/.

2.10.2.1. Effects of Volcanoes

The Hawaiian volcanoes and the rift volcanoes of Iceland and east Africa produce lava unlike other volcanoes which do not. When a lot of eruptions occur, if the output it composed of mostly lava it moves very slowly and despite having the ability to cause a lot of destruction people are not usually killed by it with as a result of slow movement. Volcano products mostly has the ejection of ash, pumice, and other fragments Called pyroclastic debris. It is ejected from by mount saint Helen and Mount Vesuvius. In volcanic ash eruptions there is production of rock fragments and natural gases such as carbon IV oxide, Carbon I oxide and hydrogen sulfide are introduced to the atmosphere as they are blown to the air and can sometimes reach the stratosphere. The ash can also cover and spread widely and even up to thousands of square kilometers. These explosions can travel even faster than sound. It may reach up to temperatures of a 1000°C and up to speeds of 62 kilometers every hour and it incinerates everything it meets in its path.

The secondary effects of volcano eruption include fires and mudslides. An example would be if mount Rainer, which lies 150 miles in the south eastern of Seattle were to erupt even gently the glaciers covering it would melt and the result would be mud, rock, and ice slide has the possibility of burying the city of Orting in Washington in a period of less than an hour.

2.10.3. Seismic Sea Waves/Tsunamis

Tectonic plates cause seismic sea waves indirectly and are commonly referred to as tsunamis. When a major seismic tremor shifts the crust underneath a water body, the energy resulting from the tremor is transferred to the fluid that surrounds the crust. The energy produced leaves and spreads out from its original site and travels through the water in wave form. Earthquakes produce tsunamis especially those that appear within the epicenter of a water body. A tsunami waves poses only a little danger while the ocean is open but can be disastrous if the tsunami waves reach the shores of the water body. Tsunamis can move up to a speed of 300 to 500 meters per hour in an open ocean. The trough of the great wave reaches the land first, which is seen as the pulling away from the shore. Disastrous consequences result when the wave hits peak. Depending on the location of the original tremor, the configuration of the local sea floor and the distance from the tremor, the tsunami varies in size, the number of waves, and the time of arrival. An example is the tsunami of 2004 that occurred within the Indian Ocean that resulted from an earth quake on the sea floor near Indonesia (Babar, 2007).

90% of tsunamis occur in the Pacific Ocean and early warning systems have been put in place since the mid-1960s however, the tsunami of Indian Ocean provided the need for an early warning system. Coastlines found where there is gradual rise of the floor of the sea are less likely to experience violent tsunamis in comparison to places that deep water surrounding them since much of the tsunamis energy is dissipated before it reaches the shore.

2.10.3.1. Effects of Tsunamis

Tsunamis with violent waves can cause a lot of destruction by crushing buildings and ships. When tsunamis are gentle, they can simply float buildings off their foundations. In the case of violent waves, if the backlash is also violent it can also cause more destruction, the damage includes loss of lives, uprooting of trees, flooding, foundations of buildings and structures are undermined and consequently food insecurity and has an economic impact.

2.11. CLIMATE CHANGE

Climate change result in new weather patterns and occurs when there are changes in the climatic system of the earth which can be adopted for a given period of time. Events of extreme weather can be a result of the ongoing climate change. The length of time goes from a short to periods or may last up to a few decades or even millions of years. Scientists have been able to record different episodes of changes in climate in relation to the earth's geological history and also with industrialization climate have also been affected by human activities which lead to global warming. Other factors leading to climate include solar radiation, a change in the earth's natural climatic cycle and human activities among other factors. Some of the signs of climate change include the trapping of more heat in the atmosphere, droughts also become longer, and more extreme, tropical storms have also become more severe as result of warm ocean water temperatures. With the rise in temperature there, also less snow pack on mountains and snow melts in polar areas. There is also a link between climate and land use. Land cover which is shaped by land use of practices which affects the greenhouse gases concentration. Despite the change in land use being an important driver of changes in climate, a change in climate can also lead to changes in land use and cover (Dufresne, 2019).

Many environmental disasters have been linked to climate change. The disasters have claimed more than 600,000 lives in the last two decades. The frequency of such disasters and their magnitude increase with time and may not subside even with the reduction of greenhouse emissions. An average of 335 weather related disasters between 2005 and 2014 have been recorded by the UN office for disaster risk reduction (DRR). This is an increase of 14% in comparison to the period between 1995 and 2015. Similarly, according to the report 4.1 billion people were affected by weather related disasters for the period 1995–2015. Most of the deaths can be accounted for by weather related natural disasters such as storms and floods. 89% of the deaths also occurred in low and middle income countries.164, 000 people have lost their lives as a result of extreme temperatures as a result of global warming. 148,000 of these deaths were caused by heat waves. The first decade of the 21st century has seen 3496 natural disasters as a result of floods, storms, droughts, and heat waves.

The global mean temperature continues to rise every year. With an increase in temperature, globally there is the likelihood of drought incidences and storms are more likely to occur. Storms may develop if there is more water available for evaporation. If more, there is more heat present in the atmosphere and surface temperatures of oceans become warmer the result is an increase in the speed of wind in tropical storms. The rising sea levels expose locations that are higher to the power of the sea especially to areas that are not usually exposed to this and also subject these areas to erosive waves and currents. Other significant impacts of climate change include reduced access to fresh water, it also leads to food insecurity, it leads to displacement and migration of families as the aftermath of disasters related to climate, pest, and diseases and health problems (Dufresne, 2019).

2.12. MINING

Mining is the process of extracting minerals or geological materials within the surface of the earth or the earth crust. Products from mining include coal, metals, limestone, potash, clay, and many others. The goal of mining is to obtain minerals that would otherwise nit be found on the surface of the earth. They are different types of mining and they include surface mining, underground mining, in-situ mining, and placer mining. There are various causes for carrying out mining they include income growth, to meet the increasing demand for resources, technological advancements, industrialization, and so many more.

Some of the most publicized environmental disasters are associated with the mining industry. Both natural and mining related causes can lead to environmental disasters. Mineral deposits from the mine that come into contact with rain water lead to the formation of acidic drainage which can cause damage to ecosystems within the vicinity of the mining area as it may cause the pollution of water bodies and harm to wildlife. Ore mining and processing can accentuate and accelerate natural processes.

2.12.1. Long Term and Short Term Impacts of Mining to the Environment

Short term impacts of mining include injuries during work or in the event a mine collapses, it can also lead to death as some people may remain buried in the mines in the event it collapses. Injuries can be caused while carrying out maintaining or repairing activities. Some people end up sustaining long term injuries that can affect their day to day life. It can also lead to hearing problems as some of the machines tend to be very loud. It is a problem in countries with low safety standards and also if the employers do not care about the welfare of the employees. And may eventually lead to damages that are not reversible. Heavy metal contamination can be a health risk to those working within a mine. Other health hazards such as cancer can be caused by mining especially lung cancer as a result of inhaling substances that are toxic which is a byproduct of the process of mining. Other than causing cancer, it can also lead to other respiratory diseases especially to workers within the mines especially if protective precautions are not taken such as using protective masks and other protection measures to ensure the safety of the employees (Dufresne, 2019).

The process of mining process involves the extraction of minerals and other resources beneath the earth surface and can lead to erosion. If the land is eroded, it can end up being in a state in which it can no longer be useful or suitable for other human activities such as agriculture or settlement. The soil in the area becomes unstable and poses further danger and further erosion. Air pollution is as a result of mining activities. Huge machines are used in the mining process to make work easy for the mine workers but at the same time, the activities produce or emit gases which are harmful including greenhouse gases like carbon IV oxide and methane. They cause air pollution and the impact is the implied consequences.

Mining activities also lead to soil pollution. In the extraction process of minerals. In addition to the material, harmful substances are also extracted.

When the substances are released in the process of mining and go on to contaminate the surrounding soil. This can also occur when the products of mining are not treated or disposed of properly they are likely to contaminate soil (Dufresne, 2019) (Figure 2.7).

Figure 2.7: Pollution of ground and surface water from acid mine drainage and contaminated dust and soil from mine dumps have exposing South Africans to high concentrations of heavy metals and radiation.

Source: http://www.mining.com/south-africa-has-failed-to-protect-locals-from-gold-mine-pollution-harvard-report/.

Mining increases the acidity of water in streams, increased sediment loads including metals; it can also spread dust with windborne pathogens and also cause the release of toxic substances. These toxic substances can be contained in ores and waste rock piles which can be derived from reactions of ore processing. They can contain cyanide and lead which may be transported by water to other regions away from the site of mining and also wind and goes further to pollute soils, ground water, rivers, and the atmosphere. Remediation cannot be used in the management of the contamination because of the scale and intensity. The effects of mining depend on how severe it is and how distorted the surrounding has become or whether there is degradation of the environment around the mining site. In a lot of cases, the effects of mining can be minimized by proper planning by creating a comprehensive mining plan including the stage for remediation and reclamation (Dufresne, 2019) (Figure 2.8).

Figure 2.8: Air pollution caused by mining activities.

Source: https://palawan.wordpress.com/2011/03/21/mining-and-its-effects-on-the-environment/.

Mining also affects animals as it can lead to the loss of habitats as the mining areas are close to the homes of wild animals. This drives animals from their environment in which they are adapted to and they may not be able to survive outside the said environment especially if they do not find a good and suitable environment. Animals can also be affected by contamination of mine by products that may be toxic. The substances include mercury and can reduce the life expectancy of the animals. It also affects plants that require a certain pH level for proper growth. The pH level is altered by mining activities that affect soil PH by polluting it. The loss of soil fertility can lead to the reduction of production and also lead to extinction of plant species or loss of biodiversity. Mining activities can also affect aquatic life when they reach water bodies as a result of improper disposal by companies that carry out mining. Aquatic life, both flora and fauna are affected by the toxic substances produced. This can be radioactive substances or heavy metals. Mining also leads to deforestation in order to meet the demand for resources. When expanding their mining territories mining companies cut down trees that include forests. Deforestation is a global issue and it has harmful effects to the environment. It also contributes to global warming and destruction of forest cover.

There are several mining disasters that have occurred around the globe especially in 2010 from Montreal, West Virginia to Yuzhuo china and as a result, a lot of people perished while working. These disasters brought to light unsafe working conditions in different mines. In America to address these safety issues, the Mine Safety Health Administration (MSHA) developed an inspection program that was targeted especially mining companies that had troubling safety record (Dufresne, 2019).

In 2010, on April 5th a mine in Montcoal, West Virginia exploded killing 29 out of the 31 mine workers at the site. It was the worst mine disaster in the United States after the Finley disaster in 1970 which was a result of high methane levels which led to the explosion.

In the same year in May, an explosion also occurred in Russia in an underground coal mine due to the buildup of methane. The mining company had a history of violating safety regulations and the explosion led to the death of four miners. Another explosion had previously led to the death of 66 people and another 110 were killed in 2007 by a methane blast.

In china in October 2010, 37 men were killed by a gas leak explosion at a coal mine in Henan province.239 miners were lucky enough to escape the incident. The report from the explosion showed that the gas concentration was 40 times more than the levels that were acceptable.

Most of the disasters that have occurred are as a result of not conforming or adhering to safety regulations related to mining and lack of proper monitoring systems especially the gas levels that lead to most of the explosions.

2.12.2. Solutions to Mining Problems

Government regulations-government regulations are important measures that have the potential to deal with the adverse effects caused by mining activities. Authorities have the responsibility to ensure that mining companies carry out mining in a manner that is responsible. This means adhering to safety measures, protecting their employees, and also preserving plants and wildlife that exist within the vicinity of the mine. In most countries, there is strict regulation while in some especially developing countries there are very low safety standards which need to be looked into. Changes need to be made in order to mitigate mining effects.

Reduce consumption-overconsumption of mining products contribute to mining problems. It will lead to the pressure of producing more products and using more resources. In order to sustain resources we require using resources in moderation (Weil, 2004).

Recycling and reusing-people should reuse stuff that is old and by doing so we can use resources more efficiently. If we find no use to old stuff, we can recycle them in the most effective way by simply separating the waste generated.

Natural reserves-to protect and preserve the environment from adverse mining effects. The government and other environmental organizations can take part in the creation of nature reserves to protect plants and animals from the harmful effects. These reserves are important to provide a safe habitat for growth and reproduction of animals. It should also provide natural living conditions and increase their chances of survival. The nature reserves also function to mitigate the adverse effects of mining operations.

Education-education is a measure against the adverse effects of mining on the environmental systems. Education should be provided to show people how mining can be harmful despite having economic benefits. The education can begin at an early age and also adults can be engaged in the process to make it more sustainable (Weil, 2004).

2.13. DROUGHT AS A NATURAL DISASTER WITH ENVIRONMENTAL IMPACT

The lack of precipitation results to shortage of water which leads to drought. Drought affects the water cycle by causing the shortage of water, damage to crops and reduction of stream flows and depletion of ground water. Drought takes places when evaporation exceeds precipitation for a certain period of time. Famine can be a result of drought that leads to starvation and death as a result of food shortage. This occurs in many parts of the world but especially developing nations. It is generally associated with semiarid and arid conditions but it can occur in areas where there is adequate rainfall and atmospheric moisture. Lack of water affects human activities such as agricultural production, industrial activities among others and this affects how people gain access to food (United Nations, 2010) (Figure 2.9).

Figure 2.9: Desertification occurring as a result of drought.

Source: https://www.thoughtco.com/drought-causes-stages-and-problems-1434940.

Drought cases occur globally, the most affected populations come from Africa and this picture gives the idea that drought is a problem of Africans. The truth is that cases of drought have been experienced in regions in the world that are arid and semi-arid and even in zones that experience temperate climates. Other than the African Sahel regions, other regions of the world that experience drought include Ethiopia, Chile, Brazil, the Philippines, India, and Bolivia. Countries such as the United States and the Soviet Union experience near drought conditions that affects food supply and demand internationally as they are grain producing regions. Current trends in the occurrence of famine and drought show that they are becoming more frequent especially in regions that border deserts and also in areas where people practice agriculture, lumbering, and grazing of livestock.

2.13.1. Causes of Famine and Drought

- Low levels of precipitation-when there is a lack of precipitation the resultant effect is drought. Lower than the average occurrence of rainfall over a period of time can lead to dry soils that cannot support agriculture. Other climatic disturbances such as high temperatures and changes in the patterns of wind lead to reduced precipitation in the area or region. It also occurs in areas that are susceptible to low precipitation.

- El Niño's-affects different parts of the world and is associated with climatic conditions that are hotter and drier. Such events are associated with droughts and occur at different times and are dependent on strength and duration of the weather events. Such events influence patterns of weather in Africa, Southeast Asia, Central America, and some parts of the United States.
- Dry season-they occur commonly in areas that experience long and dry seasons. Levels of humidity are low in such times and evaporations are high. Water bodies dry up eventually and agricultural crops that depend on water sources for support cannot be sustained and up dying. The high temperatures that are present in the dry season further worsen the drought.
- Climate change-climate change is induced by global warming is one of the recent causes of droughts. As much as climate change can increase precipitation as a result of melting ice caps and higher rates of evaporation, it can cause dryness in other areas.
- Human activities-depletion of water resources through irrigation and over cultivation can lead to drought. Other human activities that lead to drought include deforestation.

Some solutions to the drought problem include preparing for drought especially in areas susceptible to drought, desalination of water from oceans, employing irrigation techniques that are efficient, harvesting rain water during periods of precipitation, planting of drought resistant crops, use of solar pumps, reforestation, and afforestation and recycling of organic waste.

2.13.2. Environmental Effects of Drought

The environment is affected by drought in different ways. Plans and animals rely on water for survival similarly to people. The habitat and food supply are affected by drought. It may be temporary or go for long periods of time. If the drought goes for long periods of time, it may have a long lasting impact on the environment. They include fish and wildlife destruction, lack of fresh water sources that can lead to public health hazards, wildlife migration; extinction of endangered species, wild fires, poor soil quality, and wind and water soil erosion. It also leads to famine which can lead to malnutrition and if prolonged can lead to death.

Drought also has social impacts including the health and safety of people. This includes health, conflicts, and public safety. The impacts include anxiety or depression especially in relation to economic losses, health problems, loss of human life; public safety from fires; reduced income, migration, and decreased recreation activities.

Economic impacts include money lose especially for farmers, the timber industry is affected in the case of fires, power companies are also affected with water shortages, ship have a problem navigating and high prices for food and other supplies (Smith, 2003).

2.14. ENVIRONMENTAL DEGRADATION

Environmental degradation exacerbates many environmental disasters. For example deforestation leads to rapid run offs in the event of precipitation hence making a contribution to flooding. When human beings destroy mangrove forests and swamps, the coast lines ability to resist tropical winds and storm surges is decreased. Drought conditions, the severity, and the duration of the drought are related to phenomena that are natural. Conditions of drought are exacerbated by cropping patterns that are poorly done, overgrazing, and the stripping of top soil, techniques of conservation that have a poor nature, depletion of surface and ground water supply and the extent of urbanization.

2.15. POVERTY

One of the most important single influences on the impact of disasters is poverty. Factors leading to disasters would be reduced if the affected population were also not limited. Most of the studies that have been done show that those most affected by disasters are poor countries in comparison to nations that are developed and who also tend to recover faster from disasters. Poverty increases the vulnerability of people to disasters. Poverty can be useful in explaining why people in the urban poor are forced to live on hills or terrains that are prone to landslides or why they settle near volcanoes or near river banks that are prone to flooding. Poverty can also be important in explaining why droughts will mostly affect farmers that are poor rather than the wealthy. It also explains why the most affected by famine are the poor who lack the purchasing power to buy food in the event they cannot get produce from their farm. Poverty also explains why many people are forced to move across the border in order to survive. Migration that is induced by crisis can pose challenges in getting access to assistance or for long term development.

2.16. LACK OF AWARENESS AND INFORMATION

Disasters occur when people vulnerable to disasters are not aware that their activities can harm them or they are in a settlement prone to environmental disasters, or they did not know how to escape harm or take protective measures. The ignorance of the people may not always be a result of poverty but can be a function of lack of awareness of the measures that can be taken to prevent disasters, building of safe structures on safe locations, and or placing of safe evacuation routes or procedures. Some populations may also lack an early warning system and emergency preparedness measures to deal with disasters. The lack of knowledge or awareness should not be used as a justification for ignoring the coping mechanisms of the populations affected by disasters. In disaster-prone areas, there is an understanding of the threats and responses. This should be incorporated into efforts to provide assistance externally.

3

Consequences of Environmental Disaster

CONTENTS

3.1. INTRODUCTION

Environmental or ecological disasters are cataclysmic and ruinous events that occur on the environmental scale due to human activity or natural phenomena. They include natural disasters which are major adverse events that result from the earth's natural processes for example floods, tornadoes, hurricanes, and earthquakes. They have long-lasting effects that are widespread that cause human, material, economic, and environmental loss. The saddening fact is that the affected community and the society at large are incapable of dealing with it using its resources alone. Disasters can often be complex with no direct single root cause and they are more prevalent and difficult to handle in the Global South. A simple example of this is an offshore earthquake that causes a tsunami resulting in floods. Environmental Disasters can be divided into two kinds and this is based on the cause.

Natural disasters occur as a geological terrestrial process and may cause loss of life and property and further on loss of livelihoods and economic disruption. They include the aforementioned earthquakes floods but include tsunamis cyclones blazers landslides and volcanic eruptions. Natural disasters are characteristically slow or acute. This makes them easy to predict because of their geographical clustering. Natural hazards are a way of life; unpreventable and inevitable. Although to some extent they can be controlled for example the way buildings in Japan are built to be earthquake proof, natural disasters can have long-term effects (Dufresne, 2019).

Human-made (sometimes called technological or anthropogenic) disasters can also bring about environmental destruction and include the likes of industrial accidents nuclear explosions terrorist attacks and oil spills. Technological or man-made disasters like bioterrorism and nuclear accidents are caused by the interactions between human beings and nature which unfortunately can aggravate the effects exponentially (Dufresne, 2019).

The consequences, whether the disaster was natural or technological can be unfathomable despite efforts by the government and the global community on mitigating risks and raising awareness. In 1987, the United Nations General Assembly had its 42nd session declaring the 90s as the international decade for human disaster reduction. This was passed as a resolution in 1989 and the concept behind it was to curb the unacceptable rising levels of loss which disasters continue to impart on the world through combined international actions. The efforts were intended to reach four goals:

- Increase each countries mitigation capacities on the effects of natural disaster in an and effective expeditious fashion paying special attention to developing countries by establishing warning system and disaster resistant structures
- Implement appropriate strategies for the application of scientific and technical knowledge that already exists.
- Fostering the technological angle in science and engineering that would help in the reduction of loss due to environmental disasters
- Developing evaluation systems specific two locations where disasters would happen and the nature of the disaster in themselves. Measures include assessing predicting preventing and mitigating the consequences of Environmental natural disasters true technological programs and education.

Despite these efforts, that has been a growing vulnerability even still to natural and technological hazards. The downfall has been the lack of planning to achieve sustainable development or and more so short-sighted developmental activities. Despite the fact that developmental decisions are made daily on local national and international levels they are unplanned two words environmental disaster risk aversion and more often than not they increased the vulnerability to these disasters (Dufresne, 2019).

3.2. ENVIRONMENTAL DISASTER FACTS AND STATISTICS

- The global South with low and lower-middle-economic countries suffer disproportionately. 81% of the 4.4 billion lives lost lived in the Global south where only 33% of the environmental disasters actually happened.
- Women and children are the most vulnerable and the most affected are those living below the poverty line.
- The top 5 recorded countries worst hits by environmental disasters are the Philippines, Indonesia, India the United States and China in order. 8 Out of 10 of the world's most at risk cities are in the Philippines
- The rate of Environmental disasters is increasing. The annual average weather-related disasters between 2005 and 2014 was 335 that shows a 14% increase of the previous nine years (1995

to 2004) and almost double that of the decade before (1985 to 1995) (Figure 3.1).

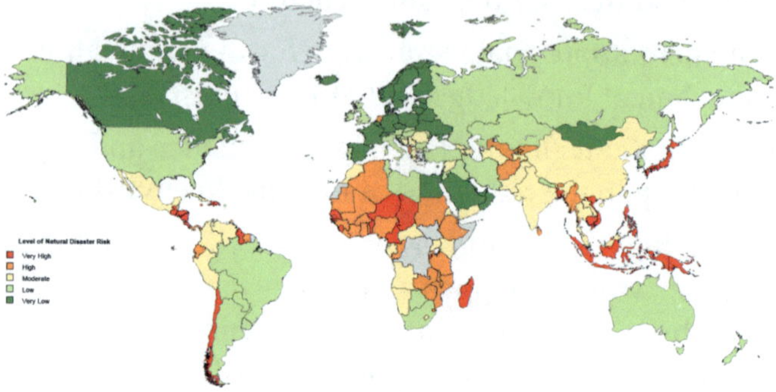

Figure 3.1: Map showing the level of environmental disaster risk.

Source: https://www.childfund.org/Content/NewsDetail/2147489272/.

3.3. ECOLOGICAL CONSEQUENCES OF ENVIRONMENTAL DISASTERS

Due to the fact, environmental disasters happen on an ecological platform it is important to first recognize the impact environmental disasters have on the environment itself. In actual sense, the end of an environmental disaster is often just the beginning. The nature of impact can vary depending on the environmental disaster itself for example many earthquakes have but minor impact on the ecosystem, floods may even be able to rejuvenate vegetation in the area, but a nuclear meltdown will leave the area uninhabitable for hundreds if not thousands of years. The reality is environmental costs are difficult to translate economically but development policy makers should devote more attention on the effects disasters have on the environment (Elliott and Hsu, 2016).

Because many environmental disasters are natural in nature, they often have mixed outcomes for the environment another benefit some parts of the natural system while destroying others. A simple example is during a blizzard the high velocity cold winds that would cause parts of the tree to thin out will eventually lead to strong tree development. On the onset, however impacts of disasters on the environment and overall development manifold. They bring an ecological imbalance and degradation in both short-term and long-term. In the Urban environment the effects are more negative and long-lasting because natural resources like water and clean air becomes scarce (when there's a fire in LA for example) the human relation to the environment actually exacerbates the negative environmental consequences. Deforestation and agriculture systems for example will make a typhoon or a landslide much more destructive in nature.

This has brought to the attention implementing environmental management systems that a comprehensive and play a major role in the reduction of risk and mitigation of effects. This should be done on both the urban environment on a human scale and the broader ecology (Elliott and Hsu, 2016).

A good case study to offer would be the hurricane season of 2017 where hurricane Harvey dropped 27 trillion gallons of rain over a six-day period. Hurricane Harvey was closely followed by Hurricane Irma that rattled the Caribbean calling people in Puerto Rico and the Virgin Islands to lose power for weeks. The natural ecosystem was greatly damaged where coastal native shellfish fish, insects, birds, and mammal habitats destroyed further exacerbated by pollutants flowing from industrial areas into other water sources and the oceans. Wildfires are also particularly bad because they can totally defoliate the natural landscape killing wildlife the habitats and their food sources. In areas where endangered species exist, they are particularly vulnerable (Elliott and Hsu, 2016).

Technological disasters appear to have a more adverse effect on the environment in that they are harder to contain and clean up there after. Chernobyl is a great example. In 1986 its reactors failed and the nuclear power plant exploded and radioactive ushering down on nearby towns and villages killing many from radiation exposure. The fire from the explosion band for nine days straight and everybody in the area was immediately evacuated. Officially, 31 people died but researcher's estimated perhaps even 10,000 people were affected. The difference here and we Hiroshima is that an atomic bomb is meant to release all its energy at once and the radiation the space is fairly rapidly but Chernobyl has scattered element with highly radioactive pieces of the reactors still buried in the soil and continues to leak into the surrounding environment. Wildlife samples collected from certain areas prove that it is not safe for any sort of living organism.

Hazardous material spills are also particularly dangerous because they are localized. Oil spills suffocate fish and even those that sit at the bottom of the water develop liver disease. The oil gets caught in the feathers of birds stopping them from flying and blocks light from reaching plants in the water halting photosynthesis and killing them. It is important to note that although environmental disasters have anime store on ecosystems they also bring increased attention and awareness into these habitats that are now vulnerable and threatened (Elliott and Hsu, 2016) (Figure 3.2).

Figure 3.2: Bird covered in oil after Exxon Valdez oil spill.

Source: https://sludge.wordpress.com/2006/08/25/remember-alaska-and-exx-on-valdez/.

The estimated initial death toll of the Exxon Valdez oil spill included 250,000 seabirds, 2,800 sea otters, 300 harbor seals, 250 bald eagles, up to 22 killer whales, billions of salmon and herring eggs, and other intertidal plants and animals. Some injured species are still recovering.

3.4. LOSS OF LIFE AS A CONSEQUENCE OF ENVIRONMENTAL DISASTERS

Perhaps the most daunting effects of environmental disasters are the loss of human life. Droughts and floods are the deadliest environmental disaster and though hurricanes and earthquakes seen more dangerous they are short lived and the long-lasting effects of floods and droughts are much more difficult to contain. In the US, floods kill more than 100 people a year. The biggest natural disaster since 1900 was the China floods of 1931 that is estimated to have taken the lives of up to 4 million people. The next top killer was the Bangladesh Bhola cyclone of 1970 that killed around half a million people. By year, 2017 2018 and 2019 hurricane Maria in Puerto Rico the Sulawesi earthquake and tsunami in Indonesia and cyclone Idai in Mozambique Zimbabwe and Malawi counted as the deadliest natural disasters (James and Paton, 2015).

The reasons why floods and more particularly flash floods are so dangerous is the speed in which they come. They have the ability to catch even the most prepared, unaware. In the US if one hears of a flash flood they would instinctively get into their car and evacuate the area but that can be a terrible idea. The Center for Disease Control (CDC) state that more than half of flood-related deaths in the US involve drivers attempting to drive over moving water. It only takes 15 centimeters of flowing flood water to knock someone off their feet and carry them away and as little as 30 cm can move entire cars (James and Paton, 2015).

Earthquakes count as the second deadliest environmental disasters and this is due to their unpredictability. The deadliest earthquake ever recorded was in 1556 during the Ming Dynasty. The Shaanxi earthquake killed a record 830,000 people. The second deadliest earthquake ever recorded was in recent memory when it lost 316,000 people. This shows that even with technological advances (over four centuries) there has been little to no effort made towards development and disaster management. Earthquakes are most dangerous if you are caught in the wrong place. The collapse entire buildings and set off landslides. If or when they happen off show, they cause tsunamis that destroy entire coastlines and swallow Islands and the inhabitants. The recent past scientists have been able to build earthquake-proof structures. Retrofitting a building to make it earthquake proof can cost anywhere between $60,000 to $130,000.

Other natural and anthropogenic disasters cause the loss of life and some of the rarer ones include heat waves and limnic eruptions. By definition, heat waves are measured in reference to the normal temperatures of the season. Heat waves are becoming more frequent and they're identified by periods of excessively high temperatures and hot weather sometimes accompanied by high humidity (James and Paton, 2015).

Limnic eruptions however are a lot more uncommon. Also known as Lake Overturn, limnic eruptions occur when carbon dioxide suddenly erupts from deep lakes. The cloud of gas suffocates everything from wildlife livestock and human beings. A violent limnic eruption can cause tsunamis as the rising carbon dioxide displaces the surrounding water. In Lake Nyos, in western Cameroon, such an eruption happened overnight and because carbon dioxide is denser than air it sinks and displays the breathable air killing nearly everyone as far as 25 km are we from the lake itself. Majority of the victims most likely woke up gasping for air which accelerated the asphyxia and turned their body fluids acidic. Meaning of found with your skin can attend and blisters on the skin caused by pressure and low oxygen levels. Local livestock and wild animals operation and the only life that survived were vegetation.

The biggest reason that disaster management systems and development policies should be paid close attention to particularly is the loss of life. It is the most distressing consequence of an environmental disaster (James and Paton, 2015).

3.5. SOCIO-ECONOMIC EFFECTS OF ENVIRONMENTAL DISASTERS

Environmental disasters are catastrophic and calamitous wreaking havoc on people's property. The negative economic effects of an environmental disaster are orphan taken into account by major disaster databases that quote damage to property and cost of repairs. The further economic impacts are often overlooked. In California, recent wildfires have seen residents lose their homes businesses and all the worldly possessions. Alabama Georgia and Mississippi residence have the highest risk of losing their home to a tornado earthquake or hurricane according to The World Bank and the National Oceanic Atmospheric Agency. The financial effects of these disasters are felt to this day and in human terms they are immeasurable (Elliott and Hsu, 2016).

Japan has been particularly hard hit on an economic level and hosts the top two most economically devastating disasters with the earthquake and tsunami of 2011 in Sunday that was responsible for an estimated mind shattering 235 billion dollars in damage. The disasters took the lives of 86490 people and until today over 13,000 people are still missing. One of the biggest reasons why the disaster was so economically taxing was that the earthquake then resulted in a tsunami which damages the cooling system of the nuclear power plant in Fukushima. The state of affairs was devastating.

In 1995, the Great Hanshin earthquake in Kobe was voted the deadliest environmental catastrophe of the year killing 6500 people and economic implications was in the tune of 100 billion dollars. Japan suffered but was able to recover fairly quickly, returning almost to normal operating levels of industry in a little over a year (Elliott and Hsu, 2016).

Hurricane Katrina (2005), California's Northridge earthquake (1994) and China's Sichuan earthquake (2008) make the list with a combined 150 billion dollars in loss of property and damages. Hurricane Katrina was especially notable in the way it damages the insurance industry costing them 41 billion dollars. This was a largest lost the insurance industry ever suffered through a single event.

There have been several studies done to examine the relationship between disaster and poverty and unfortunately, there is a direct correlation. It is estimated by the Year 2040 they could be up to 325 million extremely poor people living in areas exposed to the full range of both natural and technological hazards. This is especially alarming because as the Earth resources continue to be mismanaged by the growing population environmental disasters will be ever increasing in both number and intensity exponentially exacerbating the issue (Elliott and Hsu, 2016).

Only financial level losses are means directly during environmental disasters through such things as income loss, loss of assets and infrastructure but directly a country can be brought to its knees from the loss of tax revenue insurance losses (like the above-stated hurricane Katrina) loss of labor and loss of tourism. This ends up perpetuating the poverty levels (Figure 3.3).

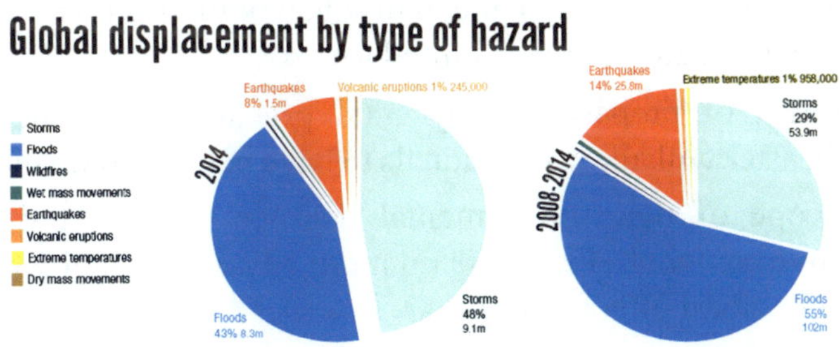

Figure 3.3: Global displacement by type of environmental disaster.

Source: https://www.internal-displacement.org/publications/global-estimates-2015-people-displaced-by-disasters.

3.6. INDIVIDUAL CONSEQUENCES OF ENVIRONMENTAL DISASTERS

Ecological destruction, loss of life and damaged to property and livelihood are the biggest consequences that environmental disasters have. Beyond the consequence of death and loss of everything around you the social implications that environmental disasters have is unimaginable online personally experienced. The nature in which an environmental disaster occurs will pressure setting reactions by persons involved. All the side effects of the disaster will offer traumatic stimuli that will impact personal and societal reactions. On an individual level, this may vary but it can be suddenly agreed that the stress levels this period of time is astronomically high. Individual physiological symptoms will include shock, nausea, vomiting, excessive, and sweating, tremors, and twitches, muscle aches, dizziness, and headaches. It is most important that seeks medical attention and more immediately so if the proximity to the environmental disaster was very close (Geipel, 2012).

Beyond physiological reactions, there are cognitive emotional and behavioral effects that will instinctively occur with the environmental disaster and these include hyper-vigilance substance abuse irritability depression relief anxiety memory loss confusion concentration problems and reduced attention span and some may be unwilling to leave the scene of calamity.

These reactions are all normal and to be expected and the first responders to the scene should be able to offer not only physiological but mental health services. The nature in which the individual will react is dependent on several factors including but not limited to:

- Level of preparedness (mental preparation of the impending disaster will allow individuals to have better coping mechanisms).
- Scope of the environmental disaster (the most damaging an environmental disaster is no more adverse mental emotional and behavioral effect to be).
- Personal loss and injury (individuals who lose their loved ones and everything that they have may never fully recover from the environmental disaster).
- General traumatic stimuli (exposure to traumatic stimuli especially for children can have irreparable effects on individuals).
- Time of the occurrence (disasters that occur in the Dead of the

night have proven to have worst mental effects than those that happen during the day).

- Environment after the disaster occurs (the surrounding area for the individuals after the disaster occurs is crucial for recovery. If the individual is kept in a polluted humid unbearably cold or hot environment, it will exacerbate the individual's reactions).
- Other factors that may determine how an individual will cope with the effects of the environmental disaster are as a result of pre-existing conditions. This is personal and may vary from person to person but include:
  - **Disabilities:** A person suffering from physical disabilities or somebody who may be blind deaf or unable to speak will experience higher stress levels than a healthy human being
  - **Health and Medical Problems:** It is evident that those suffering from diabetes or mental problems like anxiety will have a difficulty receiving medication and therefore suffer more than others who are healthier
  - **Lack of Support:** Divorced or widowed individuals will have higher stress levels after an environmental disaster than those who have a strong support system
  - **Cultural Barriers:** Those who suffer from language barriers or a generally foreigners will have a difficulty recovering from the environmental disaster than those who are locals
  - **Demographics:** Age and sex a huge Factor to the coping mechanisms of an individual after an environmental disaster. The extremely young and extremely old will have more difficulties and women tend to have a harder time dealing with the stresses that come with an environmental disaster and its effects. Almost always women or girls will suffer from the negative effects and even marital stress has been found to be exacerbated by environmental disasters.

There are guidelines offered that can helping one cope with the stress of an environmental disaster and individuals are advised to:

- Return to a routine and stick to it as much as possible;
- Reduce or overall limit exposure to the media that is covering the event;

- Staying healthy by exercising;
- Confiding in someone that you trust and keeping a positive perspective;
- Being open and talking to others about what you were going through;
- Be patient to yourself your loved ones on each other;
- Sick counseling and other expert therapeutic methods.

3.7. PUBLIC HEALTH CONSEQUENCES OF ENVIRONMENTAL DISASTERS

There is an obvious and immediate danger posed by environmental disasters but the secondary effects can be just as damaging. In relation to public health and safety the biggest and most shocking effects to survivors of Environmental disasters is the large amount of displacement to the population. When a city is rattled by an earthquake or a tsunami many people have to abandon their homes and literally start their life all over again. Displacement of population due to environmental damage is not a new phenomenon at all and historically people have always had to move when the area they inhabit has been degraded and cannot sustain their lives. The more recent phenomenon however is the potential for the displaced population like refugees for example to have a negative effect on the availability of resources and they can disrupt accessibility to healthcare education and even more basic needs like food supplies and clean water. Nearly 30 million people have to flee from their homes every decade due to environmental disasters floods and storms account for over 80% of these disasters induced displacement (James and Paton, 2015).

Healthcare instantly becomes a huge concern after environmental disasters strike. Diseases like measles cholera and acute respiratory infections are some of the most widespread and common diseases especially and displaced populations after a disaster. Severe flooding brings with its stagnant water that hosts waterborne bacteria and is a suitable breeding ground for malaria-carrying mosquitoes. Most diseases associated with an environmental disaster can be prevented by adequate interventions like simply ensuring the basic life-saving needs of the population are met initially under disaster happens immediate maternity does not depend on a specific age range however later on the mortality rate associated with the disaster is disproportionately higher among the youngest and the oldest people.

The most vulnerable groups in children especially those displaced from their families' women who are pregnant, disabled individuals and the elderly. These vulnerable groups because of setting physiological characteristics r more vulnerable to the release of toxic substances and the overcrowding but comes with displacement or populations. In developing countries especially, the five leading causes of death due to environmental disaster apart from the disaster itself are diarrheal diseases and dehydration measles malaria acute respiratory infections and malnutrition. The predisposing environmental conditions that are developed include destruction of food sources and sanitary services income loss overcrowding lack of adequate water and loss of shelter however the discontinuation of healthcare services is the major cause that would bring about the aforementioned effect (James and Paton, 2015).

Thc Essential Fmergency Relief Services as defined by the World Health Organization appropriate responses based on parlicular thing is a nice disaster in tribute lamented in the community the first step is to do a rapid assessment of the emergency situation on the affected population on the assessment should accurately define the needs and equipment the limited resources so that they are efficiently and effectively used specially to maximize life savings and other vital goals. There is a large element of specificity here on geography economics political system nature of the environmental disaster itself.

The second Step includes providing shelter and clothing because exposure to climatic conditions especially to the vulnerable can increase the caloric requirements of a person and in dire situations leads to death. The shelter should be appropriate for the climate being experience and rather than building large camps or temporary settlements, it is more recommended to rebuild within the community using the direct resources that the community has. This is a lot harder said than done especially in the Global South while limited resources are already the norm of the day however is the first step is done appropriately and assessment is accurate then it will be easier to mitigate the straining of resources (James and Paton, 2015).

Once Shelter and clothing are provided, essential elementary sanitation and clean water is the next step with an estimated minimum requirement of 3 to 5 per person per day. This is only drinking water but 15 to 20 are recommended for all of the needs that include cooking cleaning and washing. Community-level re-establishing these supplies as soon as possible and effectively can be the difference between life and death and allows public health services to be more effective. Sanitary services include a latrine and there should be at least one for every 20 individuals.

Because of the environmental disasters, disrupt health services the next logical step for man is to provide a diarrhea control program because it is a predictable outcome of disasters. The way to do this is to implement community-based education so that appropriate sanitation measures are taken to prevent diarrheal diseases. Healthcare centers would be strained immediately environmental disaster workers however; they should anticipate these needs and also cater for dehydration using appropriate low-cost strategies. Dysentery and cholera or almost part and parcel of an environmental disaster and should be planned for appropriately (James and Paton, 2015).

The next disease and source of healthcare problems is measles and malnutrition that tends to be the major source of mortality in displaced populations that are especially crowded. Vitamin E deficiency is very common and actually contributes to measles related mortalities. At community national and international level, agencies should cooperate to ensure appropriate distribution of measles immunization or vitamin A supplements. At the very basic community level, health officers must be able to assess the cold chain as part of the health care assessment.

Establishing reproductive health and HIV services is a great way to improve on the primary medical services as beyond the crude mortality rate there will be deaths due to preventable causes in the following weeks and months after the environmental disaster. Which include setting up emergency obstetric and neonatal care while measures for the prevention and management of STIs. Rape sexual violence and exploitation can be widespread in displacement and refugee camps following an environmental disaster although the extent in which that is recognized is very limited. This again is disproportionately an effect of the global South with widely varying estimates of the numbers of victims reported. Civil unrest and general who are part of the human made or technological disasters that are part of the Environmental calamities and this is where sexual and reproductive health risks are highest. Sexual violence and rape is recognized today as almost an integral feature in Wars and conflicts but also come along with other natural disasters as well.

Public health organizations should also feature disease surveillance and health information systems in the environmental disaster management programs as they are necessary in measuring the level of effectiveness that they have and allows them to also reassign and allocate priorities. This again should be done at a community level where healthcare professionals also share this information with higher-level authorities on the national level (United Nations, 2010).

The organization of human resources is another mobilization effort that could make public health emergency management services function optimally and indirectly save lives. The initial shock of an environmental disaster can paralyze the population that is affected disabling them from effectively responding in a quick and organized fashion. Having a predetermined and predefined set of Rules and an emergency plan can provide the community with an ability to cope until external health and Resources arrive. On a community-level this happens by conducting rescue operations and having a management system able to do so, organizing food and water distribution planning health services, corpses, and gravesite management as well as conducting assessments on the various necessities (United Nations, 2010).

Coordinated activities on a larger scale also enable relief efforts to be more effective and this ultimately depends on the quality of assessments made of the affected communities so that priorities established.

3.8. MENTAL CONSEQUENCES OF ENVIRONMENTAL DISASTERS

An environmental disaster can have largely traumatic effects on children especially. Dealing with a large amount of Destruction and confronted with the death of friends and loved ones the development of post-traumatic stress disorder and other psychological conditions can be an after effect. Psychological effects of Environmental disasters include a combination of shock and denial. In some, this can be extremely paralyzing especially in children and those with pre-existing mental conditions making it challenging to take the required steps of picking up the pieces like assessing property damage and loss and finding temporary housing. The shock eventually fades away but it's replaced by much stronger feelings that can hit even months after the disaster has hit.

High levels of stress experienced by everyone involved including those people who come from outside in to help. For residence the safety and refuge, that they called home is destroyed and great feelings of insecurity can bring nightmares and anxiety. Extreme stress is common but when it persists for months after the aftermath it is most likely a sign of post-traumatic stress disorder. People suffering from PTSD from a storm that swept their homes can experience panic attacks or flashbacks when even light rains with thunder and lightning occur. Post-traumatic stress disorder interferes with the ability of a person to control the emotions on the onset can be delayed from one week and up to three years after events. In Japan, for example research has been done into psychiatric assistance for Disaster victims and in the US, epidemiological studies state that at least half the population experiences a traumatic event in their life and so PTSD is therefore not an uncommon thing. It's important to Note that natural disasters can be particularly traumatizing in an insidious fashion because the post-traumatic stress disorder is experienced by large amount of people at the same time. According to the American Psychological Association common symptoms of post-traumatic stress disorder and trauma in general are:

- Feelings are increasingly intensified and unpredictable with mood swings anxiety and depression as common manifestations.
- Flashbacks that are vivid and repeated manifesting Dreams and Nightmares sometimes and leads to apparent physical reactions.
- Sleep and appetite issues.
- Confusion and the difficulty in making decisions.
- Fear of events repeating and anything similar thereof.
- Personality changes especially in relation to interpersonal skills. One may become more avoidant or hostile.

Generally, survivors of Environmental disasters are recommended to seek professional psychological or psychiatric help more especially so if they're unable to return back to normal functioning. The first stage of post-traumatic stress disorder is characterized by hyper-vigilance and adrenaline surges that tamper with normal functioning of the neuro-transmitters. This can persist up to one month but if it lasts longer than that, the Victim is considered. To invented second stage which is characterized by a sense of inability to control a situation and general helplessness. Some body functions such as breathing heart rate and blood flow become a regular especially in light of events or anything associated with. Personality changes like those listed above occur here. The third and final stage she's the victim display low spirits and dejection with depression and anxiety overwhelming the individual. Anyone involved with the environmental disaster even remotely is vulnerable to stress disorders including Healthcare and rescue workers but the risks are higher with the proximity experienced (Geipel, 2012).

3.9. TYPES OF ENVIRONMENTAL DISASTERS AND THEIR CONSEQUENCES

Of course, the nature of an environmental disaster in and of itself present a different set of circumstances and consequences to deal with and it is important to see the distinctive factors that come into play after certain types of natural and technological disasters occur. The above-listed loss of life, economic losses environmental destruction, and health factors weather physical or mental r general and encapsulate effects of all environmental disasters. Continuing further, we will see some more unique consequences. These are natural disasters that have environmental consequences such as droughts and environmental disasters, which are caused through human activity.

- Avalanche;
- Blizzard;
- Earthquake;
- Wildfire;
- Floods;
- Freezing rain;
- Heat waves;
- Hurricanes;
- Landslides;
- Lightning strikes;
- Limnic eruptions;
- Tornadoes;
- Tsunamis;
- Volcanic eruptions;
- Bioterrorism;
- Civil unrest and war;
- Urban fires;
- Hazardous material spills;
- Nuclear radiation;
- Power failures.

3.9.1. Avalanches and Their Consequences

An avalanche is a snow version of a landslide and sometimes is aptly called a snow slide. It occurs when large slab of snow is lying on top or a weaker layer of snow which fractures off and slides down a slope. They are primarily composed of blowing snow and their butt the bigger they get the larger the capability of sweeping and picking up momentum of ice rocks trees and other surface materials. They are ordinarily classed by size destructive potential and the nature in which they were initiated. They can be triggered naturally the loading of new snow or rain but they also have more artificial triggers like explosives or skiers that overload the snowpack (Dufresne, 2019).

Globally averages of nearly 200 people die every year from an avalanche with people who are caught in them dying from suffocation, trauma, or hyperthermia. They are not a rare occurrence and anywhere there is a mountain with a standing snowpack and Avalanche can occur especially during winter or spring but really can happen anytime of the year. They can end up damaging utilities and communication system destroying pipelines that carry gas or oil causing them to spill or destroying power lines causing disruptions in electricity for thousands of people.

The Andes mountain range that stretches across the western side of South America is where the deadliest landslides would naturally occur and this is again due to lack of preventative and mitigative measures. The deadliest the deadliest ever recorded landslide killed over 20,000 people in Huascaran-Ancash, Peru. It is the worst natural disaster ever to strike the country and it occurred on May 31st 1970. It was triggered by an earthquake 21 miles off the coast of Peru that destabilized the snow on the northern rules of Mount Huascaran. Large chunks of ice snow and mud spanning 900 m wide and 1.6 km long cascaded down the mountain at speeds of up to 300 km per hour (Dufresne, 2019).

Apart from the damage to life and Property together with the social economic and health, impacts Avalanche is also pose the risk of coming together with flash floods. They also changed the entire ecosystem changing weather patterns causing crops to fail in the lower fields (Dufresne, 2019).

3.9.1.1. Management of the Disaster

The preventative measures that can be taken especially where Avalanches pose a significant threat and these include repeatedly traveling on a snow park where the snow has accumulated and grooming the area with the machine as well. Grooming ordinarily happens in ski resorts and it's the process of manipulating the snow for recreational purposes with the use of a tractor or a snowmobile. Explosives are also used extensively to trigger smaller avalanches that will break down the parks that are unstable.

Artificial barriers have also proven to be effective in reducing Avalanche damage and mitigation can come in several forms including snow nets that are flexible snow supporting structures built out of Steel or nylon cables. It is also it is also not unheard of to build Earth Mounds along the path which the Avalanche would slide down to slow or overall stop the flow of destruction. Technologies like high-resolution cameras or motion sensors can act as early Warning systems and can detect Avalanches even several days in advance. Modern radar like interferometric radars act as complex alarm systems that detect the Avalanche emergency services to close roads or evacuate the lower-lying areas and one such technology is employed in Zermatt Switzerland (Figure 3.4).

Figure 3.4: Avalanche path with an 800-meter vertical fall in the Glacier Peak wilderness, Washington State.

Source: https://commons.wikimedia.org/wiki/Avalanche_chute.

3.10. BLIZZARDS AND SNOWSTORMS AND CONSEQUENCES

Blizzards are severe winter storms have high winds lasting for a prolonged period of time. They bring with them large amounts of ice snow and dangerously low temperatures. The low temperatures that come with them prompted the National Weather service in the United States to implements something known as a wind chill temperature index. Wind chill wind chill is the calculation of how cold it actually feels rather than how cold it actually is. Blizzards can shut down cities entirely sometimes even four days trapping people in their homes. Due to heavy winds and snow electrical cables can be Fell leaving people without electricity and the most severe Blizzard can make transportation impossible (Babar, 2007).

A little bit more dangerous than avalanches, blizzards cause roughly 400 deaths per year and have the exact same consequences as Avalanches with a different severity. The deadliest blizzard ever recorded happened in Iran in the winter of 1972. The severe storm lasted a week and resulted in the death of approximately 4000 people. Up to three meters of snow was dumped all over majority of the country and Southern Iran especially received as much as 8 meters of snow burying victims. Rural villages of Kakkan and Kumar had 0 survivals. Whole communities were buried alive.

3.10.1. Management of the Disaster

Disaster plans for blizzards include insulation of your homes I'm paying attention to weather reports. Supplies should be gathered to plan ahead for several days' even weeks without power. Other individual other individual measures include an emergency supply kit for your vehicle and signing up to the local Communities warning systems. If caught outside the first and most obvious action would be to find shelter right away and staying warm.

3.11. EARTHQUAKES AND THEIR CONSEQUENCES

Earth tremors occur when seismic activity occurring in the interior of the Earth producer a shaking on the surface. The range in size from those that are very weak and on felt too violent one that can propel objects and even people into the air. Earthquakes can occur pretty much anywhere on the planet and they count for one of the most severe of environmental disasters. Ordinarily they coincide along tectonic plates with the most active being the circum-pacific belt. This stretches from New Zealand and South up and across New Guinea and Japan all the way up to the Alaskan west coast cycling back down north and South America. About 80% of the energy below tectonic plates is released around this belt. Earthquakes are more often associated as a natural event but can also be man-made caused by the injections of fluids into deep wells or the detonation of underground explosions. Reservoir induction and nuclear explosions together with tectonics and volcanism encompass all the causes of earthquakes and they can happen and they can have their epicenter on land or in the sea (Babar, 2007).

Ground shaking is the most familiar effect or earth tremors and they can be vertical or horizontal shifting the ground by raising it dropping it to tilting it. The geomorphological changes vary from landslides, mudflows, to significantly damaging buildings railways and other structures. There is more evident damage on soft and unconsolidated ground (sometimes liquefying sandy terrain) rather than hard rock and some can be felt as far as 100 km from the source. It is uncommon for them to be felt underground with reports of considerable damage directly above mines which were unaffected. They are accompanied by sounds and lights that are very distinctive; the lights coming from electric induction along the source of the earthquake and the sound being likened to an underground train.

Earthquakes are particularly dangerous because the after effects are more often than not other environmental disasters in themselves. On land on land, they can cause mudslides and landslides and if they happen on the seabed, they can cause very long wavelengths of water and seismic sea waves known as tsunamis. They will they will propagate along the open ocean but when reaching the shallows they will crest and their amplitude increases, growing in height and power and can be extremely destructive. In a nutshell, effects of earthquakes include:

- Ground shaking;
- Surface faulting;
- Ground failure;
- Flow failures;
- Landslides;
- Tsunamis and Sieches.

The deadliest ever recorded earthquake happened almost 500 years ago in Shaanxi China that killed 830,000 people across 97 counties of the

Ming Dynasty. The reason the earthquake proved so deadly is because of Loess cave settlements that were particularly common during this period that became the final resting place for many thousands. More recently an earthquake in Haiti, 2010, killed 316,000 people although the figures are still disputed (Babar, 2007) (Figure 3.5).

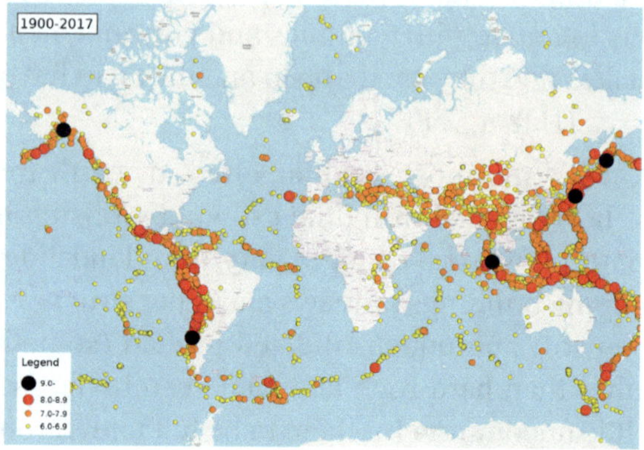

Figure 3.5: Major earthquakes in the 20th century, legend depicts measurements on the Richter scale.

Source: https://en.wikipedia.org/wiki/Earthquake#/media/File:Map_of_earthquakes_1900-.svg.

3.11.1. Management of the Disaster

Seismographs and accelerometers are used to measure earthquakes and their ground motion. The earliest of mechanical seismographs had a pendulum with allows mass up to several tones to swing when seismic activity occurs and scratch lines on a rotating drum. In modern times, technological advancement has allowed us to have more precise pendulum seismometers. Due to the fact that 70% of the Earth is covered by water, there has been a great need to have instruments that can record seismic activity on the seafloor and Japan has spearheaded this technology and has had a seismograph system on the seafloor off the Pacific Coast that was installed in late 70s. All in all the last half of the 20th century has provided various technological advances and mobilization of efforts especially with the worldwide standardized seismographic network (Smith, 2003).

In relation to observing and predicting earthquakes there has developed a new interdisciplinary branch of engineering to minimize the risk of structures being damaged overall loss of property and lives. Earthquakes cause deaths to secondary effects like landslides tsunamis and such however, the greatest loss of both lives and Property comes from the collapse of man-made structures. Therefore, the most effective way to minimize the damage that earthquakes cause from an engineering perspective is to construct structures that are capable of withstanding seismic activities. Buildings nowadays can even be seismically retrofitted where existing structures are modified to make them more resistant to ground motion. Earthquake insurance is a way to mitigate financial losses for building owners against losses resulting from them and has developed into a large industry of property insurance. On both a national and global scale there are emergency management strategies that can be employed to consolidate risks and prepare for the eventual chaos and consequences that follows an earthquake.

For individuals' preparedness, measures should also be taken by securing large items around the house that could injure someone like water heaters and education should be widespread to those living in earthquake-prone areas so that they know exactly what to do when the Earth was shaking. People who also live near large bodies of water that are earthquake prone means the earthquake preparedness is paired with tsunami preparedness as well (Smith, 2003).

3.12. WILDFIRES AND URBAN FIRES THEIR CONSEQUENCES

Wildfires or rural fires can consume very large areas of combustible vegetation. Depending on the vegetation, present wildfires can be classified as bushfires desert fires forest fires grass fires and they are ordinarily unplanned and uncontrollable. The three major causes from a natural perspective are dry climate, lightning, and volcanic eruptions. From a human perspective, causes include arson, power line arcs, and discarded cigarettes. These causes vary depending on which part of the world you are in. In Canada and China, for example lightning is the biggest cause of wife eyes while another part of the world human involvement and activities like agriculture and land conversion burning are the major causes. Wildfires spread in accordance to the amount of oxygen and flammable material present (Weil, 2004).

One of the more notable wildfires is the Peshtigo fire of 1871 in Wisconsin. It's actually occurred it's actually occurred on the same day as a great Chicago Fire but it is largely forgotten despite the fact that it killed far more people. It is estimated that it took the lives of as many as 2500 people though the estimates are difficult to confirm due to many records burning up in the five South. The fire spread over half a million acres and is still up-to-date the deadliest wildfire in American history. Its ignition is still under speculation and has developed many different theories but because of the fact that it happened at the same time as the Chicago fires and even fires in Michigan it is presumed that the impact Fragments of the Biela Comet may have been a reason. Other more contentious ignition sources include small files that were burning in the area for land clearing operations after the very dry summer of that year. The fire grew so big and so fast it carried an intensity that created its own wind systems and continue to sustain it. Super-heated flames of up to 2000°F moving on winds up to 110 miles an hour where created and burned over 1.2 million acres (Weil, 2004).

3.12.1. Management of the Disaster

Wildfire prevention is the most effective method of reducing the risk of fires as well as minimizing the severity. Prevention measures consider equality logical balances and protection of resources and as much as American firefighting policies allowed naturally caused fires to burn out on an ecological scale it is important to consider the human role played in wildfires. In Europe for example 95% of the forest fires that occur there are human-caused. The most famous Wildfire prevention measures were initiated by Franklin D Roosevelt many of us can remember Smokey bear and his admonition "only you can prevent forest fires" (Figure 3.6).

Figure 3.6: Smokey bear.

Source: https://www.williamsnews.com/news/2019/aug/06/history-fire-prevention-smokey-bear-way/.

3.13. FLOODS AND THEIR CONSEQUENCES

Floods are simply overflow in water in an area that is ordinarily dry and how one of the key areas of study in the discipline of hydrology and they are an important area of concern in agriculture civil engineering and public health. Floods are probably the most dangerous environmental disaster to deal with and occur when water bodies (from the smallest being rivers and streams two very large lakes and oceans even) overflow. They can happen very slowly over time why is the most dangerous normally occurring very quickly without even visible signs of rain. They can be caused by upslope factors when the areas upstream or uphill shed water down below Al down slope factors ultimately include factors like elevation and flow channel restrictions. Weather conditions that can exacerbate the intensity of a flood includes unusually warm rainfall melting snow packs or ice natural activities like the breaking of beaver dams or earthquakes causing tsunamis (United Nations, 2010).

They're both primary and long-term and more permanent secondary effects. Floods can sweep away whole communities contributing to an immense loss of life and Property Damaging Power Transmission and generation and contaminating water supplies causing water Borne diseases like typhoid cryptosporidium cholera and many others. Because roads canals and transport systems and infrastructure in general can be destroyed, it may be difficult to get help to those affected. They can destroy farmlands and make the land unworkable damaging the food resource for the community depends on. This contributes to the further effects of economic hardships and even those houses that were damaged could have moved from the water which result in adverse health effects more specifically respiratory illnesses. The economic implications are even more heartfelt in urban areas.

Six of the 10 deadliest recorded floods orchid in China topping the list was the 1931 China floods that cute up to an estimated 4 million people. Majority of the people who died here died as a result of farming and disease that was caused by The Flood. The preceding 3 years so a very long drought and the harsh winter of 1930 created large deposits of snow and ice in the mountain areas that melted later in 1931 and flooded the Yangtze during the heavy spring rains. The spring rains were followed by extreme cyclonic activity during the summer and in July, alone 9 cyclones hit the region. Meteorological stations record 600 mm of rain reported by four different ways of stations along the Yangtze River. The following autumn did not offer any reprieve as heavy rains continued to pour in the region and Many Rivers did not return to their normal levels until November. 53 million people were affected by those rains and the further economic shocks will so damage there was barely any foods with some people having to sell their children to actually survive and grim cases of cannibalism also reported. The most lethal effect that the Flood hard was the various diseases that came afterwards that included cholera measles malaria dysentery and schistosomiasis (United Nations, 2010).

3.13.1. Management of the Disaster

Flood safety planning especially that list by the National Weather service that is a Federal Government body in the US states that the best thing to do during a flood is to turn around and don't drown. This is basically a statement saying that when a flood does happen, the best course of action is to get out of the affected area and not try and cross it. The best defense is most likely to seek higher ground. In relation to urban planning essential service areas like emergency operations centers and hospitals, also police, fire, and rescue services should be planned to be built in areas least at risk from flooding. Other engineering considerations includes short-term monitoring like forecasting warning and response operations and designing structures to control flooding by observing previous and presented floods areas and their Flood Heights. In many areas across the world, there are waterways that are prone to flooding that are monitored and managed well we detention basins reservoirs and levees built to prevent overflowing of their Banks. Seawalls and barrier Islands are also building various places in the US and Europe as defense measures. Urban planning measures include building more porous areas that allows water to drain better especially services like parking lots and expanding man-made so as to accommodate storm water and flood water (United Nations, 2010).

Flood focusing and anticipating floods before they happen can be a life-saving caution and this is done to computer models that are studied together with hydrography. Because modern technology allows us very accurate weather forecasting abilities especially with the use of radar good quality data can be used to predict a flood well ahead of time and allow other precautionary measures to be taken.

3.14. HURRICANES AND CYCLONES AND THEIR CONSEQUENCES

The hurricane or a tropical cyclone is characterized by a low pressure Center that circulates very strong winds producing squalls and thunderstorms. If you are caught out at sea during a tropical cyclone, you will observe very heavy rain and large waves that can even cause shipwrecks and disrupt international trade and shipping. Over the last two centuries, alone hurricanes and tropical cyclones have been responsible for nearly 2 million deaths worldwide. They are similar to floods in that they create other water-related consequences like flooding and large bodies of standing water causing infections and other diseases. The deadliest ever recorded hurricane was the great Galveston hurricane known in the region of the Great Storm of 1900 that tour of the storm of Texas destroyed 7000 buildings in the area. The city of 30,000 habitants the city of 30,000 habitants so 10,000 people left homeless and many estimates up to 8000 fatalities occurred

3.14.1. Management of the Disaster

A unique consequence of hurricanes and tropical storms is in the way that it is able to contaminate the whole area by mixing water systems with stored chemicals sewage human remains asbestos lead and so on. This requires the first this requires the first responders to be particularly careful. Other than, that preparedness and response actions to be taken during a hurricane is monitoring the news because tropical storms are almost impossible to miss and Hurricanes come from a long way out. If advised to evacuate the area one must do with the utmost urgency.

3.15. TORNADOES AND THEIR CONSEQUENCES

Tornadoes sometimes known as twisters or whirlwinds are meteorological phenomena that are characterized by rapid rotating column of air between the surface of the Earth and a cloud above. The major the major effect is wind damage similar to hurricanes but innocence best dramatic the beauty is that they can also be detected and rigorous attempts have been made by the US to create efficient warning systems. A major misconception is that they cannot occur in city but the Salt Lake City tornado of 1999 disproves that misconception among several others. Structures that are built around to needle prone areas are required to have basements or bunkers where people can take shelter in case they are not able to evacuate the area. There are insurance companies that are in charge of property damage in case of tornadoes and Hurricanes alike.

3.16. HEAT WAVES AND THEIR CONSEQUENCES

Heat waves are characterized by periods of excessively hot weather humidity and is ordinarily measured in relation to what the climate pattern for that area demonstrate they are relatively new phenomena and have been more credited to global warming and climate change. They can also be caused they can also be caused by aforementioned wildfires They can be especially dangerous to the elderly and young children and those with pre-existing conditions and prescription medications like diuretics anti psychosis and anti-hypertensive. They can cause:

- Heat rashes that are maculopapular that inflame and block the sweat glands;
- Heat cramps that are extremely painful involuntary spasms of muscle groups from intense exertion;
- Heat-syncope that is fainting and dizziness as a result of the body overheating;
- Heat exhaustion and heat stroke classified as hyperthermia;
- Heat edema which is the swelling of the body as a reaction to try and retain fluids.

The only management protocol that can be taken when a heat wave strikes is to stay cool which means staying indoors and using air conditioning and all possible methods to keep the body temperature low.

3.17. VOLCANIC ERUPTIONS AND THEIR CONSEQUENCES

A volcanic eruption spews lava tephra ash and many other things into the surrounding environment. Like earthquakes they also depend on activities going on below the surface of the Earth and the deadliest ever recorded was in Tambora Indonesia in 1815. 80,000 people lost their 80,000 people lost their lives interest of the Year Without a summer. Because they depend on seismic activity as well, they can be measured and predicted in kind almost the same way that earthquakes are. The easiest preventative measure to take is to evacuate the area whenever national and international safety organizations predict a threat (Babar, 2007).

3.18. NUCLEAR RADIATION ACCIDENTS THEIR CONSEQUENCES

Nuclear radiation accidents are defined by the atomic energy agency as in I went to the house let the significant consequence to people environment to the facility the most famous being those of Chernobyl and Fukushima. Fukushima nuclear reactor disaster was unique in that it was caused by an earthquake first which caused the tsunami that damaged the active reactors. Radiation leaked into the atmosphere and evacuation protocol had to be extended just beyond the affected area of the tsunami itself up to 20 km away from the area. Large amounts of water were contaminated with large amount of water was contaminated with isotopes that are Radioactive and released into the Pacific Ocean which are the tremendous impact in the ecological sense. Generally, the main concern during a nuclear accident is that of radiation itself and how it affects the people around.

In conclusion, human beings have long thought that the creativity ingenuity and technological powers could meet the challenges of disasters both natural and man-made. He continues to be he continues to be tested just especially as the Earth's resources continue to be eroded by growing populations and the demand for these resources. It is sad to say but it is likely that environmental disasters will increase in both number and quantity overtime and the Communities affected by these catastrophes vanity appropriate coping mechanisms and this begins by first understanding the consequences of Environmental disasters.

4

Factors of Vulnerability to Environmental Disasters

CONTENTS

4.1. INTRODUCTION

An environmental disaster is regarded as a catastrophic outcome with relation to the environment as a result of human activity. This sets it apart from the concept and phenomenon known as natural disasters which comes as a result of natural tragedies of nature. In this scenario, the results of human activities alter the ecosystem which leads to a wide and timeless range of consequences. These catastrophic events have severe and detrimental effects on the ecosystem which are often in short duration, and thereafter have a long-lasting impact on the agricultural field, biodiversity, human health under the economic stability of mankind.

Environmental disasters can be divided into two major categories. First, we look at natural climate also known as weather events. Some of the examples in this particular category are drought, wildfire, floods, earthquakes, tornadoes, landslides, volcanic eruptions, and tsunamis. Although the causes of natural environmental disasters have no particular involvement with human activities, in a few cases the results are worsened by the continuous work of man's hands. For instance, as a result of urban development and construction works that caused damage of the coral reefs in the Indian Ocean, the 2004 Tsunami caused a huge environmental suffering to those areas being under consistent human construction. In the places where the coral reefs we're healthy, the acted as a buffer and a stable ground which deflected and turned away the powerful giant wave. Another example we can give would be the avalanches which nest in the Philippines in the year 2006 seem to be one of the most disastrous avalanches in the history of mankind. The rockslide came down and leveled a mountain village identified as Giunsaugon to the ground and claimed the lives of thousands of people. The reason stated for this event what's the uncontrolled an untamed constant deforestation and mining activities around the village. Many people have disputed the kind of effect on the level of impact human activities pause to exacerbate natural disasters. Humanity is able to have a very great influence natural environmental outcomes some of them being floods, drought, typhoons, and many more. For instance, in the case of floods, a long period of rain, heavy rainfall in one sitting, incredible of circles or very wet soils, could be a cause of natural flooding. However, human activity such as deforestation, hydroelectric power, agricultural activities, natural wetland destruction, or urban development could lead to flooding. This simply shows us the threat humanity poses to a natural environment.

The second category of environmental disasters lists causes which occur through human involvement. Some of the examples I can list is chemical spills, oil spills, and nuclear incidences all of which cannot take place without human activity. In addition, terrorism activities as well as wars are very much disastrous and cause irreversible catastrophic damages to the ecosystem. In several cases, these environmental disasters caused by human activity have a longer and effect on the environment done the damage is brought by a natural environmental disaster. An example, in 1989, the enormous oil spill that occurred when the Exxon Valdez supertanker ran aground into Prince William Sound. This oil spill continues to have major environmental consequences and grieve repercussions to date. some 12 years after the oil spill, significant oil deposits found to be toxic in nature to many species (both fish and mammals), persisted throughout the neighboring affected areas. In 2002, the size of the population of at least 8 species of fish in mammals continued to severely bear the impact of the 1989 oil spill.

On various accounts, environmental disasters take a tremendous toll on the ecosystem. That being said, increased attention to the threatened habitats comes as an additional plus. In many cases, the increased attention by governmental, non-governmental, and intergovernmental organizations results in legislation which leads to a direct reduction of the impact of future environmental disasters. With reference to the Exxon Valdez Oil spill, strict regulations were put in place to cub and hinder oil spills in the future as well as much stronger regulations put in place foe oil shipping industries. With reference to the 2004 tsunami, the United Nations began an intervention and organized The Indian Ocean tsunami warning system that alerts citizens in the event of another gigantic wave is heading towards land (Figure 4.1).

Figure 4.1: Image of the 2004 Indian ocean Tsunami aftermath.

Source: https://ichef.bbci.co.uk/news/768/cpsprodpb/505D/production/_110237502_gettyimages-51895841.jpg.

With human involvement and increase in activities that continued to jeopardize the environment, disaster risk continues to increase. Disasters are more than often seen as acts of nature, but there is not an accurate translation of reality. The major factors that influence and fuel disaster risks Are contributions of human and societal vulnerability combined with the capacity to respond to and or reduce the impact of environmental disasters. A disaster can be defined as a critical disruption of the normal functioning of a society which causes a widespread material, economic, environmental, and human loss which goes beyond the ability of the community or the society affected to cope using their own resources. Due to the rampant increase of environmental disasters. Disaster risk reduction (DRR) is a concept that has come up in an effort to reduce the risk factors that lead to environmental disasters. DRR can be defined as a framework of several elements concerned with the possibilities of minimizing vulnerabilities and disaster risk throughout affected communities and largely the world to prevent or hinder the great impact of hazards within the wide context of susceptible development. Environmental degradation, livelihood choices, behavior, and patterns of settlement all greatly contribute to disaster risk (Birkmann, 2007).

4.2. VULNERABILITY

Vulnerability can be described as a serious danger deeply rooted in various conditions of social physical economic an environmental exposure which all need serious assessment and to be managed on a continuous basis. Looking at the interaction of vulnerability elements in a society is very important when one is seeking to understand the risk reduction and it should also be addressed with reference to the balance of the factors that come into play (Birkmann, 2007). The following is a list of factors that should be considered carefully when we attempt to examine and estimate the extent of vulnerability in a community:

- **Physical Factors:** These are normally material oriented, which come from the land use and planning, architecture, and the field of engineering. vulnerability from a physical point of view although still viewed as continually being widened in scope, still mainly concentrates on the consideration of construction location and the building environment. Physical vulnerability is also often determined by various aspects some of them being population density levels, design, and materials used for constructions infrastructure and housing, as well as the remoteness of settlements and sites.

- **Social Vulnerability:** This is normally connected to the level of individual well-being, community's well-being and societal well-being. the level of literacy, education, the states of security and existing peace, the access to basic human needs and right, social equality, positive traditional values, their customs and ideological beliefs, evidence of good governance, among many other characteristics form a collective organizational system. There are groups or communities that are more vulnerable than others, and people who are unfortunately less privileged do you too their social class or their position in an ethnic minority, the young and the old in age might also be worse off. Public health which house is physical, mental, and physiological well-being it seen to be a crucial aspect of social vulnerability. Traditional knowledge systems paired width cultural aspects such as indigenous beliefs, tradition, and ways of coping with impact are very important dictators of the trajectory in risk preparation.
- **Environmental Vulnerability:** This includes the extent of distribution of natural resources depletion and the data on resource degradation and erosion. lack of access to clean air, safe, and clean water as well as sanitation and inappropriate ways of waste management, witnessed more in highly heavily populated areas can exacerbate the socio-economic vulnerability. Extremely poor environmental conditions for example soil degradation or growing food scarcity can very easily threaten the food security for people who depend on the produce of land forests and marine environment for those who have their livelihood there. As natural resources continued to become scarcer, the range available options two societies become more and more limited which causes a drastic reduction in the availability of coping solutions and mechanisms which also reduces the local resilience to environmental disasters. Over stretched periods of time, environmental factors can greatly increase the vulnerability by forming and you pattern of social harmony and economic stability which eventually forces migration of whole established communities.

With the coverage above, we have seen the level of human activities and the great relationship it has to environmental disasters. We have also seen the great impact human activities have on the environment. We can now go into looking at the underlying causes of environmental disasters which are mainly human oriented in relation to their activities socially, economically, politically, and ideologically. The various underlying causes we will look at are; poverty in relation to environmental disasters, limited access to power structures in relation to environmental disasters, limited access to information in relation to environmental disasters, limited access to ideologies in relation to environmental disasters, limited access to economic systems in relation to environmental disasters and finally illness and disabilities in relation to environmental disasters (Birkmann, 2007).

4.3. UNDERLYING CAUSES

4.3.1. Poverty and Environmental Disasters

This is basically defined as a state being extremely poor. Poverty and disasters are very closely linked in very many parts of the world. Poverty is mainly considered a very important factor for evaluating and measuring household sensitivity and the extent of one's ability to disasters. This is because the level of household income shows the number of peoples coping and adaptation capabilities. Before going any deeper, we have to understand what is meant by the term poverty. The most common definition is in direct relation the household per person income. An individual is unfortunately considered poor if he or she has an income level that falls below a certain minimum which is designated as the level necessary to meet various needs of this individual some of the examples being food, affordable housing, clothing, better education, affordable quality health care and so on. Currently, the minimum level which is referred to as 'the poverty line' By the World Bank ranges at US$1.25 and US$2 per day. By 2005, the top 10 to 20 poorest countries in the world are reported to have an average poverty line of $1.25 that day, which translates to about 1.4 billion people found in the developing world lived their daily lives in poverty (Birkmann, 2007).

The existing link between disaster risk and poverty is seen could be Undeniable do you to the abundance of evidence proving the same. Results from the worldwide consultations with the poor done by the World Bank in 2000 and the UNDPs in 2004 give a report on mutual links between disaster and development. It is widely recognized that the poor often settle and build a livelihood on marginal lands, in poorly constructed homes with very little or poor access to clean water and sanitation. For example, 50% of the poor in Africa live in slums many of them found in the city. People living in such conditions are seen to generate immediate unsafe conditions. Such conditions immediately classify the poor as sensitive and more prone to disasters which then aggravate their poor economic status. Given that they have limited income and also have a hindered financial saving habit, the preparedness of the poor to take action and respond or recover from a disaster is close to zero. This is due to their limited access to disaster proof technology, the ability to relocate to less hazardous areas, or replacement of lost items as well as seeking insurance for their own health and their property. In many regions of the world, the risk keeps growing do you to a great increase in population; drastic climate change, rapid urbanization, and environmental degradation are some of the driving factors to disaster risk increase for the poor people worldwide. The very weak financial structure of the poor is mainly due to their occupations.

If one thinks about it, the manpower in factories, mining fields, marine occupations are some of the occupations that the poor people are in great numbers. In places where these people are in their own pieces of land, they are pushed to logging for charcoal, overgrazing their animals for those that are nomadic in nature and overfishing which in many cases is deep sea fishing causing irreversible damage to the life in water. Deforestation, overgrazing, and a great increase in land degradation have brought serious damages to our ecosystem and have increased the risk of disasters like floods, avalanches, and landslides. In more cases, women are seen to bear the effects of disasters more than. It is also seen that more women than men are injured or face eventual death due to hurricanes and floods. This is because women are less likely to have the know-how to swim and they are also often restricted from running quickly well fast buy the types of clothing they wear. Their role as caregivers of their children and older people in the society is also a very huge restriction for them to leave their homes without having everyone by their side (Cutter, 2012) (Figure 4.2).

Figure 4.2: People try to cross a river as others look from a damaged bridge in El Salvador in the wake of tropical storm Agatha, which led to consumption per head falling by 5.5% and increased poverty by 14%. Photograph: Jose Cabezas/ AFP/Getty images.

Source: https://www.theguardian.com/business/2016/nov/14/natural-disasters-poverty-world-bank-climate-change.

Poverty puts people in a very vulnerable position. Lack of access to assistance in different environmental disasters such as floods, volcanic eruptions, avalanches, hurricanes among many other leaves the poor at a very high risk of loss of life and an eventual loss of livelihood due to the damages caused by the environmental disaster.

4.3.2. Limited Access to Power Structures and Environmental Disasters

A power structure is defined as a system that has influence between any and every individual who is part of a selected group of people. This definition or description captures the way power and or authority is delegated and shared amongst people within a particular group for instance a nation, an institution or an organization, a government, or a society. Some of these major groups are associated with business, economics, education, government, healthcare, and many other fields in the various orientations. A powerful and stable power structure may be formal and deliberately directed to maximize on values such as efficiency and fairness. Some power structures are informal and are a set of roles similar to those found within dominator hierarchy whereby the members of the social group interact and in most cases, the interactions are aggressive creating a ranking system. Power structures are also very fluid and have changes that occur constantly, be it slowly or very rapidly, at times evolving or revolutionary, and lastly peacefully or violently. We therefore sit and ask ourselves; how does the power structure affect the environment? What could put our environment at risk based off of the power structures in place?

Eluding to the power structures in many African countries, the democratic way of government allows the people to have a voice. The environment is a key component of man's day to day activities. The delegation and division of power in government ministries is key to the survival of the environment. Gatekeepers are also very important in that they keep an eye out for the best interests of the environment even in the face of the very much advancement we have in today's life especially technologically. Failure due to greed for power in the African context has brought grieve circumstances to many people who have to deal with lack of resources and proper facilities that are responsible for the well-being of the environment. Drought and hunger are some of the major problems faced by these communities caused by the poor power structures found in these areas of government. However, developed countries, seen, and regarded as better placed in the terms of mature developed power structures also have a part in which they have failed. The industrial sector and mass production of various products made for internal and external use in these countries has put the pollution levels above reversal levels. Water pollution, soil pollution and air pollution in these countries, as many as they are puts the world on its knees due to the deteriorating climate change affects greatly the environment. There are also a number of wars that have caused damages to the environment due to power structures in various supper power countries having disagreements which threaten the lives of people as well as affecting the environment (Cutter, 2012).

4.3.3. Limited Access to Resources and Environmental Disasters

The definition of resources is simply supply of money, materials, staff, and other assets that can be counted on by people or an organization in order to have smooth functionality directed towards effectiveness. They can broadly big classify according to availability into two which is renewable and non-renewable resources. Some examples of non-renewable resources coal, crude oils among many. Some renewable resources our water wind solar energy natural gas air, etc., they can also be divided as actual resources and potential resources based on the level of development and they use that they have. Looking at the basis of origin, they can be termed as abiotic and biotic. In addition, on the basis of their distribution we can turn them as localized or ubiquitous. The benefit of having resources I was having to utilize them in the situation that they are most helpful. Resource utilization may include proper functioning of a system, increased wealth, or at times enhancing well-being. from a human point of view, a natural resource can be defined as anything but can be obtained or naturally exists in the environment which satisfies the human needs and once. The concept of resources has spread underdeveloped widely across many well established and developed areas of work. We see it in action in economics, biology, computer science, Management, ecology, human resources, and many other career fields. The direct relationship of recourses to the environment is first of all-natural resources. These are said to be those derived from the environment by humans for the satisfaction of their needs and wants. Most of the natural resources found on earth are detrimental to human survival while others are used only to satisfy desire. The introduction of conservation is directed to manage natural resources aimed at sustainability (Cutter, 2012).

In the events of environmental disasters, for example the Australian wildfire that hit in 2019–2020 that claimed lives and cause a lot of damage to miles of forest down to the ground in ashes is one that could take away the livelihood of many and also their homes (Figures 4.3 and 4.4).

Figure 4.3: Record-breaking temperatures and months of severe drought have fueled a series of massive bushfires across Australia.

Source: https://www.bbc.com/news/world-australia-50951043.

Figure 4.4: The Kenya red cross responds to humanitarian emergency following deadly floods.

Source: https://media.ifrc.org/ifrc/press-release/kenya-red-cross-responds-humanitarian-emergency-following-deadly-floods/.

These kinds of disasters cause thousands and at times even hundreds of thousands of people to evict from their houses and the places they have known as home all their lives. They lose their way of life and especially economically, they suffer the most. Food, clean water, sanitation, health facilities, and medical assistance are some of the recourses that become scarce or even extinct in these types of situations. The humanitarian relief organizations step in where the government lacks in recourse provision. The manpower needed in these kinds of situations to help save lives and property is an immense number and is at times not met.

The relocation process consumes time and these people have to find new grounds to call home. Moving into these new lands causes consumption of the resources that are there and, in most cases, there are settlers there already. This limits the resources that the existing settlers had and more limited for the new settlers coming in due to environmental disaster eviction. Water borne diseases, drought, malnutrition, and hunger, exposure to new and at times harsh climate change are some but a few of the challenges added to the already existing suffering of these evicted residents (Pelling, 2012) (Figure 4.5).

Figure 4.5: Flooding kills 50, displaces thousands in East Africa.

Source: https://www.aigmediapro.tv/flooding-kills-50-displaces-thousands-in-east-africa/.

4.3.4. Limited Access to Information and Environmental Disasters

Information can be defined as the learning of facts or details about something or someone. It can also be seen at that important part that answers the questions of what things are or why they are as they are. It has different weights and interpretations based on the different contexts or notions that it is attached to. The various contexts it can be in relation to is for example; communication, data, education, form, constraint, knowledge, meaning, mental stimuli among many others. One could ask what we use information for and as rhetorical as it seems, many of us do not know. Information use is a concept with regard to understanding first of all the sources that people choose to rely on and the application which makes sense in their lives in different circumstances. Information creation may be for a different reason from the reason why people decide to use the same or different piece of information. For instance, the information provided by a radio broadcast, in the viewpoint of the show producer could be to influence the audience who are tuned in to the show. The same piece of information could however be used as a piece of commentary or to create humor by a comedian or could be used as a political tool or a propaganda spreading item. People use information to seek a meaning or deeper understanding to various situations. At times, there is the application of information instrumentally to acquire a skill or to achieve a goal. Other times it is used cognitively, which is to generate or come up with ideas (Kapur, 2010).

Information can be gathered in public hearings, meetings, television or media outlets, computers which is under the umbrella of technology (internet), lobbying, negotiations, mediations, group forums, committees, etc. The concept of access to information means an individual being in a position to look for, receive, and use the information appropriately and effectively. It at times covers Scientific, indigenous, under traditional knowledge which is also called the Freedom of Information. Access to information is a critical aspect of life that enables citizens to exercise their voice, to affectively monitor closely the activities of the government and to hold them accountable, and also to have informed dialogue about various matters that are important in society that bring a change to their lives for the better. It is seen as a crucial tool of empowerment for the citizens which include the vulnerable and marginalized people who helps them claim their rights and that which they are entitled to as citizens of that country. The identifiable success of good governance lies on a two-way street. This is the willingness of the government to be open and transparent for all to see and the ability for the citizens to use information gained to demand and ask questions, both which may be stretched in low capacity settings.

In relation to the environment, the term Environmental information comes into play. This can be defined as the information that human beings can gather about the air, the soil, the water, land, flora, energy, noise, fauna, waste, and emissions, etc., with this information, one can be in a position to understand policies, decisions made in relation to the environment and also activities that affect the environment. The availability of this information and the accessibility is what till date is a matter that has yet to be addressed. People living in remote areas having no access to technology or electricity, could be living in a lot of ignorance. They are also completely unaware for the environmental changes and challenges we face in the days we live in. even in events where the media comes out to educate the general public on environmental conservation, these efforts fall to deaf ears due to a number of factors. Some being language barrier or technological barriers. Other times we come out to educate the general public about environmental disasters and what measures to take in the event of such emergencies before they receive proper help, the same barriers apply (Kapur, 2010).

In an earlier discussion we saw the great effects of mans' hand in the environmental disasters. People having information about environmental disasters and conservation, is a big step towards sustainability. With limited access to this same information, man continues to destroy the environment in situations such as deforestation, overfishing, overgrazing, air, and water pollution which continues further environmental degradation. Avalanches, landslides, floods, desertification, oil spills, just to mention a few disasters that would strike the face of the earth would have been avoided, if only there was the access to information on the environment and environmental disasters.

4.3.5. Limited Access to Ideologies and Environmental Disasters

An ideology is said to be a set of values that is subscribed to by a group of people or an individual upheld for reasons that are not entirely relating to knowledge or to a degree of its validation. Majority of the existing fields that ideologies are most practiced are politics, economics, religion, culture, philosophy, psychology, sociology, etc. Ideologies are characterized by four basic identities:

- An ideology must have power over cognition;
- An ideology must have the capability of guiding an individual's evaluations;
- An ideology must be able to provide adequate guidance towards an action;
- An ideology must be able to stand logically coherent.

Environmentalism is a broad ideology with a lot of concern aimed at environmental conservation and protection and health improvement of the environment. The improvement of environmental health seeks to particularly merge the environment and humans, animals, plants, and non-living things. This ideology upholds the very importance of the very environment we abuse as a curtail part of the earth system such as climate as well as provide good life for all living things here on earth. Restoration and preservation of the environment has brought about social movements and a lot of organizations that focus on nothing but the welfare of the environment. This same ideology is opposed by Anti-environmentalism. This stands based on the statement that the earth is much less fragile than is portrayed by some environmentalists. This ideology also portrays environmentalism as an overreaction and hysteria to human contribution to climate change and human recourse use (Kapur, 2010) (Figure 4.6).

Figure 4.6: March against Monsanto Vancouver, Canada, 25th May 2013.

Source: https://en.wikipedia.org/wiki/March_Against_Monsanto.

With reference to environmental disasters, one is expected to be an environmentalist. The lack or limited access to these ideologies to people leaves us at the mercy of those that exploit the earth and rob humans the good climate and safe environment they deserve. With this critical aspect of environmental preservation and restoration lacking, we are bound for a down-hill ride. Environmental disasters are immanent and will keep increasing in number and intensity leaving man as the blame.

4.3.6. Limited Access to Economic Systems and Environmental Disasters

An economic system is defined as the means or ways in which the government in a country distributes its recourses and how it trades its goods and services. The control production in five aspects; labor, capital, entrepreneurship, physical, and information resources. This is the most important part as to how a society or a country stands and sustains itself. These resources that are distributed play an important role to how the country or society creates revenue as well as utilizes that which is within its boundaries. Some economic systems have an international accommodation. This is where the relationship between them and another are created and built on an economic level. On the page of the environment, the economic system is a part of dear importance. Environmental economics is but a branch of economics. This is concerned with environmental matters. This has become a widely studied subject for the reason of the growing environmental concern in the 21st century (Birkmann, 2007).

This type of economics undertakes particular issues such as the costs and benefits or alternative environmental policies that deal with solid waste, air pollution, water pollution, sanitation, toxic substances, and global warming. Some of the solutions that advocate for environmental conservation to curb environmental disasters as well as climate change are; environmental regulations, taxes, and tariffs imposed on pollution, property rights that are better defined quotas on pollution, etc. Having strong and stable environmental systems in place directly translates to the capability of dealing with environmental disasters. This means that you have resources in funding and staff that will be in a position to deal strongly and effectively with environmental disasters should they happen. The limited access to this privilege, especially in developing or third world countries is a serious disadvantage. Should a disaster strike, there are not much the citizens that live in abject poverty in rural areas with no information or resources can do to save themselves from the situation (Birkmann, 2007).

4.3.7. Illness and Disabilities and Environmental Disasters

An illness is a disease or a while of having a sickness that affects the body or the mind. A disability is a physical or mental condition that inhibits completely or limits an individual from living a normal life. Disabilities are not inherent, but they are a relational concept, a function of the relationship of the individual with their social or physical environment. Disabilities can be caused by illnesses or by accidents. Natural disasters can cause disabilities and illnesses. The same can happen with environmental disasters.

A report given by the WHO on injuries and disabilities by earthquakes and tsunami in Asia stated that injuries and drowning have accounted for more deaths so far in countries that suffer from tsunami. Environmental disasters such as floods and earthquakes cause the most damage in terms of coasts, lives claimed and also the number of those injured as a result. Some of these people lose limbs; sight, hearing, internal bleeding, and puncturing of organs and other sever damages physically. There are those that suffer mental damage from events of this nature. Flash images of the immense destruction, loss of livelihood, loss of loved ones and having to see dead bodies in their numbers can leave its mental scars on many. Majority of those that suffer mental damages are women and children. The loss of their sons and husbands could also be a huge point to the direction of psychological trauma (Figure 4.7).

Figure 4.7: Nepal earthquake: India and China send rescue teams to Himalayan Nation.

Source: https://www.wsj.com/articles/nepal-earthquake-china-sends-search-and-rescue-team-to-katmandu-1430032246.

4.4. DYNAMIC PRESSURES

Dynamic pressure and underlying causes have a huge hand in the influence of vulnerability especially in relation to environmental disasters. Dynamic pressures are consisting of social structures and processes that have a serious influence on the extent to which a community or society is vulnerable to disasters. The main structures are the people themselves, organizations that affect the community's capacities, and the processes that are the way they instill influence. We will look at some of the dynamic pressure that is related to environmental disasters (Figure 4.8).

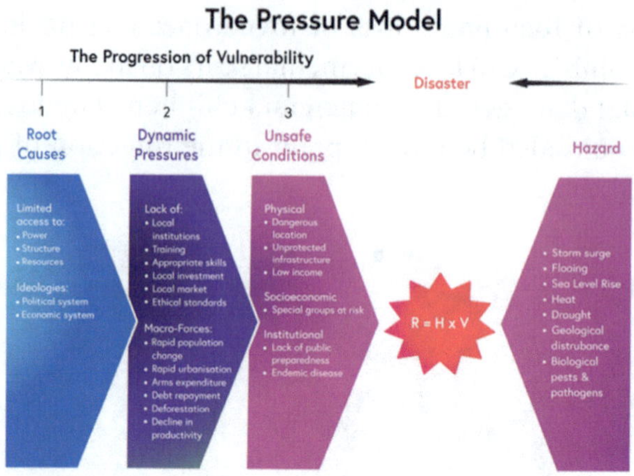

Figure 4.8: Vulnerability progression.

Source: https://www.futurelearn.com/courses/humanitarian-action-response-relief/0/steps/60984.

4.4.1. Lack of Local Institutions and Environmental Disasters

The role of local institutions cannot be undermined in a community or a society. Reports have stated that for the effectiveness of any risk management mitigation strategy depends strongly on the nature of the risk in discussion, the population of the community and the institutions characteristics, as well as the risk management capacity. The image below is a clear depiction of a disaster management cycle (Figure 4.9).

Figure 4.9: Disaster management cycle.

Source: https://www.researchgate.net/figure/The-phases-of-the-disaster-management-cycle-and-examples-of-related-activities-modified_fig1_265619198.

Certain local institutions have a vital role to play in the disaster management cycle. They, prior to the environmental disaster, offer information and educate the community on various facts and aspects of the society and its responsibility towards the environment. Them being absent, there is a fundamental gap that would cost the same community in terms of loves and property as well as financial damages, in the event of an environmental disaster.

4.4.2. Lack of Education and Environmental Disasters

The lack of education can be defined as a state of people having below average level of common knowledge on basic concepts that are vital for human day to day life. For instance, basic knowledge on mathematics, reading, and writing, etc., are part and parcel of our daily activities and relations with others. Education on environmental disasters helps the community to predict, cope, resist, and rehabilitate themselves to recovery after the disaster. Lack of it puts humanity susceptible to poverty, homelessness, unemployment, increased crime rates, and the high levels of illiteracy in a given society. In the field of environmental disasters, when areas are hit with disasters like earthquakes or floods, there is a great loss of public infrastructure. It is important to note that people with extreme vulnerability, require special training which could be unavailable due to location or in most cases due to lack of adequate funding. There are also huge rates of health complications making it hard for children to make it to schools. There are no proper roads or in the event they existed, they were destroyed by the floods or any other disaster that hit, making it almost impossible for children to take transportation to school. Other times there is a high rate of poverty induced by natural disasters and these communities have a hard time sending children to school with little or no money (Weil, 2004).

4.4.3. Lack of Training and Environmental Disasters

In environmental disasters, there is significance for emergency and disaster management training. Organizations such as the Red Cross Society, AMREF, World Health Organization, Action Aid International, etc., step in to help in these times of need. The volunteers, first response teams, and staff are trained in information on health and medical services, rescue missions and many relief programs. The importance of the training for these first responders as well as people of the society is to beef up the level of preparedness and response in the event of any occurrence of a disaster. Lacking this training for the community, leave them in the dark. In the event an environmental disaster strikes, they will have no knowledge on how to deal with injuries, infrastructure damage, rescue, life resuscitation, basic first aid, and tools that could be used or improvised to help the situation.

4.4.4. Lack of Appropriate Skills and Environmental Disasters

In relation to training, education, local institutions provided by a good economic system by the government, the acquiring of appropriate skills with regard to environmental disasters. These skills are vital before the disaster, during, and after. The whole package allows the people to be aware for their vulnerability to disasters as well as for them to have the appropriate measures that are needed for the preparation and recovery of the effects caused by the environmental disasters. Skills such as digging, first aid, lobbying for manpower, relief improve, emergency rescue services among others, help a great deal with the saving and preservation of life in the event of environmental disasters.

4.4.5. Lack of Local Investments and Environmental Disasters

Major environmental disasters cause irreversible damages to infrastructure and people's livelihood which is more often than not the economic backbone of the society. These disasters also bear long-term consequences that affect the economic growth and takes down steps towards poverty eradication. In countries of the Caribbean and Bangladesh, there is proof of a decline in sensitivity to tropical storms as well as floods. There is also proof of boosted resilience emanating from economic transformation and public positive response to disaster reduction. Countries with high risk to disasters continue to be vulnerable especially in sub-Saharan Africa. The biggest investment man can make today is DRR for those countries with high vulnerability rates and also environmental conservation and restoration.

This will allow the society to be ready in terms of funding, resources, staff, and emergency response teams in the event of the inevitable. The absence of this investment leaves us out in the dark. The increased chances of disaster risk put humanity at the mercies of the same environment being destroyed. The society would be better placed if there was DRR and management investment for the prevention of economic decline in the event an environmental disaster hit.

4.4.6. Lack of Local Markets and Environmental Disasters

The concept of a local market is the buying and selling of products and services in the same region or area which they were generated or produced. This is a key factor in the economic growth of an area. There is consumption of goods and services in the same region and none of them goes to waste due to the risk of external trading with other regions. The importance of local market is not only for the buying and selling of goods and services for those in the same radius, it also creates employment for those in the same radius. The region with a local market has a sustainable way of living and it stays strong and invulnerable to disasters. There would be services of a hospital or a clinic, a shop for homeware and food, fresh farm produce, clothing, masonry services, plumbing, among many other local convenient for day to day living for man. The lack of this local market system would mean that they would have to depend on inter-regional markets or international markets. This does not necessarily promote the community and the society in the same way that a local market would.

The lack of this system condemns the society in the event of an environmental disaster. They would be in a better position to respond quickly to the disaster. Available services and resources can be lobbied with haste to save that which can be saved. In the event of total destruction, a community of this nature and culture among them has a better chance of survival and recovery due to their work ethic or marketing culture. On the other hand, environmental disasters would bring down this very system to the ground. Destruction of property, land, loss of lives in numbers, displacement, and relocation of people would be the inevitable end for the same local market system.

4.4.7. Lack of Services and Environmental Disasters

This can be defined as the shortage, absence, or insufficiency of an item or a service that is needed in a particular situation or for a particular reason. Emergency medical care, rescue teams, evacuation teams, relief donation services for food, water, and clothing are just but a few essential services that would be needed at the time of an environmental disaster. There is the concept of environmental services that are part of the recovery and treatment of the people's condition after and during the environmental disaster. Waste-water treatment, water collection methods, drainage systems for flooded homes, waste management systems that can be given as services for those affected could be essential for their survival and recovery. The lack of these services means death in large numbers, destruction of property, loss of hope, critical medical conditions for the inured, etc., for those affected.

4.4.8. Lack of Freedom of Information and Environmental Disasters

Freedom of information is an extended arm of freedom of speech. It is one of the most fundamental human rights recognized internationally by law. It is explained as a concept of where one has a right to access information or expresses themselves on print, writing, art, and in the modern times, the internet. It clashes quite severally with the right to privacy which states that one has the right to keep their affairs unknown to others. With regard to the environment, there is the Environmental Information Regulations of 2004 which gives access to environmental information to the public which is held by public authorities. There are two ways in which this is done:

- The public authorities have to make environmental information proactively available; and
- The public audience is entitled to request for this information from the public authorities.

This means that everyone has an equal right to gain access to the information. An applicant does not require any reason as to why they are seeking this information. All requests are required to be treated equally and one must treat all information released from the public authorities under the regulations as thought the information was being released to the world at large. In the context of environmental disasters, looking at South Southeast Asia on December 26th in the year 2004, tsunamis, and earthquakes hit hard. More than 178,000 people lost their lives and around 130,000 others went missing. Loss of property, homes, and livelihood crippled the functioning state of these regions. An estimated amount of £21.5bn was the cost of the recovery stage of relief operations. There are however years of damage yet to be restored before the lives of those affected can return back to normal.

In the same scope, the right to accurate and appropriate information of the people is in need of protection and enhancement. In these scales of disasters, exercising, and fulfilling this freedom plays a huge role in ensuring that assistance is effective and available and relevant locally but on the same page, it also saves lives and helps preserve human dignity. With the freedom surprised by the public authorities due to an array of reasons and the general public being unaware of the existence of the information and their right to it leaves a lot to chance. The general public will suffer the consequences of having their rights kept from them as well as being ignorant about this information on the environment and all that is in relation to it, good or bad (Birkmann, 2007).

4.5. MACRO-FORCES

4.5.1. Population Expansion and Environmental Disasters

This can be defined as population increase or growth. This is the rise of the number of people in a given area which translates to a global total of people. The global human population increase rate is about 83 million people annually. This translates to 1.1% every year. The current global population stands at 7.774 billion people in the year 2020. There is an expected increase in population according to studies curried out worldwide and it estimates that by the mid 2030's we will be at 8.6 billion human beings on the face of the earth. The key factors that attribute to this rapid growth in population in many countries across the world are the poor standards of living, whereas the countries with low rates of population have high standards of living (Figure 4.10).

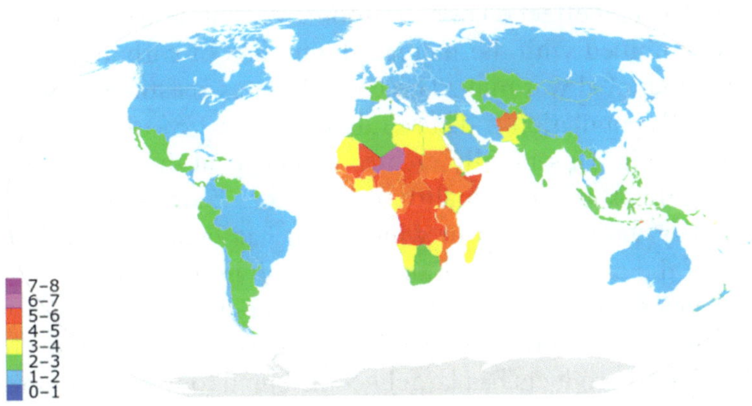

Figure 4.10: World map showing the global variation in fertility rate per woman according to CAI World Factbook's 2016 data.

Source: https://en.wikipedia.org/wiki/List_of_sovereign_states_and_dependencies_by_total_fertility_rate.

High population rates are caused by a number of factors. Poverty, poor contraception education and use, reduction of mortality rate, better medical facilities, low life expectancy, increased intensive farming, faster climate change, child labor, immigration, etc., are some of the many reasons that attribute to overpopulation. There is a possibility that a sparsely populated area to become densely populated in the event it is not able to sustain life. The growing advances in the technological world have greatly affected humanity in many ways. The ability to save lives and offer better medical services as well as creating better medical treatment for human beings is one of the biggest gift technologies has bestowed upon humanity. A direct result to this advancement is the increase in life span also called life expectancy which leads to population growth. In the history of our species as human beings, the rate of death and birth has always been at a constant balance which maintains population growth rate at a sustainable level.

Since the bubonic plague, population growth has been constantly increasing. Between the 1400s under the current 21st century times, a countless number of wars, natural calamities in man-made hazards have happened. There is however an unfortunate side of this scale. Depletion of natural resources, environmental degradation, wars, and conflicts, the rise in unemployment, high cost of living, etc., are some of the effects of overpopulation have costed humanity. The overuse of coal, oils, and natural gas has created a great effect on our environment. Emission of industrial and motor gases combined with the increased amounts of carbon dioxide into the air has led to global warming. The United States is estimated to contain 5% of the world's population and currently produces 25% of carbon dioxide emissions in the air. This rapidly creates a change in the climate patterns which greatly affects how our quality of living has been compromised. We can also identify various ways in which man has created a livelihood at the expense of the environment. Industrial revolution has been one of the biggest revenues generating aspect for many countries around the world. The manufacturing and production of tools, items especially machinery, and chemical based products has largely contributed to water pollution, soil pollution, air pollution all directed to meet the demand levels for goods and services for the large population globally (Coppola, 2010).

The resources that are available do humans as of today are not enough to sustain life over a long period of time. Water, food, and other basic necessities essential for human life and survival continue to be depleted due to the constant rise of population. In relation to environmental disasters, this remains inevitable for humanity. Pollution being one of the biggest causes of climate change and environmental degradation leaves a huge margin that cannot be reversed due to the lack of concern for environmental restoration and conservation. Population increase has a direct relationship to vulnerability and the disaster risk rates of a given area. Larger and more intense environmental disasters tend to take away more lives and do a lot more economic damage in densely populated areas. Population distribution also plays a great role in the increased vulnerability to disasters. Around 80% of the United States, for population is residents in urban areas which results do an increase in population that is concentrates in the coastal regions which are flood prone. Limited escape routes, congestion, dense infrastructure in poverty all contributes to vulnerability. In the African context, the increased number of slums in urban areas creates a never-ending possibility for environmental disasters due to poor sanitation, for infrastructure, poor population distribution, increased pollution in air and water, etc. That being said, mankind is continuously exposed to environmental disasters due to the large numbers of population. The likelihood these disasters increasing in number and intensity only continues to increase putting millions of lives at risk as well as economical vulnerability at an inevitable state (Coppola, 2010).

4.5.2. Urbanization and Environmental Disasters

Urbanization can be defined as a population shift from rural areas into urban areas, the decrease in population distribution of people living in rural areas and urban areas as well as the way in which societies accept and adapt to change. It is also mainly the way in which towns and cities come up and grew in size as more people continue to work and settle in these areas. It is also the process which explains the growing of towns and cities into urban areas as people permanently concentrate there. The United Nations does not offer its definition of the word 'urban,' but it provides a definition and description for the word 'urban place.' This means a locality that has more than 2,500 people who live in this locality. In Peru, these terms apply to an area that has a population center with 100 or more that are settled there.

After the Neolithic period which began Roughly 10,000 BCE, is when people started to form small permanent settlements. Areas with more than 10,000 people did not exist. Around the 1800s, less than 3% of the world's population lived in cities of about 20,000 people. This has since increased by one quarter all the same population dated the mid-1960s. By the 21st century, it is reported that half of the world's population already settled and resides in urban centers. With this many people living in one area, there is a need to generate a sustainable economic system double support and sustain life. Invention of various tools that boosted and made life easier for men became the order of the day bringing forth the wheel, the television, the newspaper, the telephone, agricultural tools, development of architecture, medical advancements, etc., that was also the creation of currency which is important for trade and the survival of any economic system of a given population. This supported local markets and gave birth to international and cross regional trade.

We also have the discovery of mining, which is to exhume natural resources that were found either on the surface of the earth, deep in the earth's crust, deep in water, etc., that needed construction or excavation materials and equipment which man sat down, thought about, and created (Coppola, 2010) (Figure 4.11).

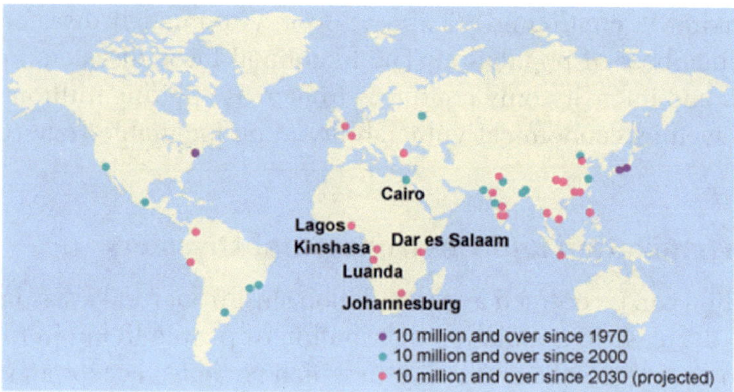

Figure 4.11: The top mega-cities in the world in 1970, 2000, and 2030.

Source: https://www.open.edu/openlearncreate/mod/oucontent/view.php?id=79940&section=3.1.

The causes of urbanization come down to two main reasons; the increase of population and the rural urban migration. The increase in population has a significant cause to the growing urban population. The increased birth rate overtime according to the demographic transition model has led to a great overpopulation within cities. This however has another devastating side to it. Homelessness, unemployment, high crime rates, increased cost of living, and poor living standards are some of the disadvantages that large populations have to deal with in urban areas. In developing countries, urban rural migration is a common misconception that is devoted to a better life. The same concept can apply to immigration from third world countries to developed countries. This migration is greatly influenced by the economic growth on technological development. It is driven by the pool and push factor that attracts people from rural countryside areas into urban areas. Employment opportunities found within cities stand as the pull factor. Many cities have industries that are located within them and offer opportunities for employment at high urban wages. This later on reflects to poor living conditions and a lack of opportunity for good paying employment in the rural areas which is the main push factor. People move away from the rural areas due to poor health facilities, limited education, as well as economic opportunities, environmental change, floods, drought, among other pressures of livelihoods in the rural areas (Babar, 2007).

The impact of urbanization has a mixed outcome. There are positive impacts as well as negative impacts that bring changes on either side of the scale. Driving towns and cities are identified by an essential element of a prosperous economy. The putting together of economic and human resources in one basket stimulates innovation and business development, science, and technology and many other industries to develop. Access to affordable quality health care, education, social services and cultural activities are more than readily available for people living in urban areas than they are available for people living in rural areas. The density of urban population makes it easier and also cost efficient for the government as it provides essential goods and services for these residents. The negative side of this coin is the poor housing that develops in these cities. in Addis Ababa, in 2008 was found to have 80% of homes built in the city we're all rated and classified as slums due to the physical state and building materials that caused a deteriorating and overcrowding effect in these areas (Babar, 2007) (Figure 4.12).

Figure 4.12: Slum in Addis Ababa.

Source: https://www.open.edu/openlearncreate/mod/oucontent/view.php?id=79940&section=5.1.

Poor housing with very small and overcrowded houses built very close to one another is one of the largest characteristics of a slam. They constantly suffer from poor sanitation, Poor waste disposal, lack of sanitation facilities and inadequate healthcare facilities are only the tip of the iceberg. Water supply is also very low as well as lack of proper drainage facilities. Solid waste is also a big hazardous problem that they continuously suffer from (Figure 4.13).

Figure 4.13: Urban rivers contain wastes from many sources.

Source: https://www.open.edu/openlearncreate/mod/oucontent/view.php?id=79940&section=5.3.

Air quality ties is also very poor due to the vehicle exhausts, domestic fire smoke, factory emissions, diesel powered generators and the increased levels of dust. Health, food is limited in supply, and high cost of living is some of the downsides of the life in urban areas. Keeping in mind all the above listed negative aspects and contributions of urbanization, there is a lot of environmental risks and hazards we are susceptible to as humans. Waterlogging and flooding, drought due to air pollution, respiratory diseases, and pandemics based on the poor sanitation and deficient healthcare facilities, earthquakes, oil spills and many other disasters that can hit humanity continue to brew slowly and marinate in front of our very eyes. The damages that would be consequent to an environmental disaster would take years of recovery. Loss of lives in their thousands based on the poor infrastructure found in these cities and the scale of architecture in the same city could kill many and cause damages that would take away the livelihood of many. The disaster risk and vulnerability brought about by urbanization in the 21st century is great and at times scary if humanity does not pay more attention to the reduction of disaster risk. There should be a response and investment in disaster management facilities as well as environmentally friendly policies that regulate the level of human activities that compromise the state of the environment (Weil, 2004).

4.5.3. Environment Degradation and Environmental Disasters

Environmental degradation can be stated as the disintegration and deterioration of the earth's environment through its assets such as water, air, land, soil; and the eradication of valuable wildlife. This is one of the biggest threats of humanity faced by the world today. The United Nations International Strategy for Disaster Reduction states that environmental degradation to lessen further the ability of the earth to meet its demands and needs. As we continue to thrive as a human race, the consequences of our actions continue to harm and corrupt the environment.

Land disturbance attributed to land damage, pollution whether in air, water, soil, land, or noise, overpopulation, landfills, deforestation, and natural causes such as tides, storms, wildfires, etc., are some of the largest contributors to environmental degradation. The impact of this on the eco-system that man clings on for existence on the earth is worrying and the worst part is that there is very little concern or eyebrows raised on the matter of environmental degradation (Weil, 2004).

An example of the Amazon 2019 wildfires is quite recent and very significant. The Amazon rainforest makes up to 60% of all the worlds' rainforests. It is said to be the lungs of the entire earth. With the human population increasing, the Amazon is able to clean our air and also regulate the rainfall all around the world. Having any damages sustained by the rainforest could jeopardizes the entire existence of man on earth. An example of deforestation, this can pose a major threat to the oxygen supply on the earth. Environmental degradation has a direct effect on water. Water quality sinks to almost zero due to pollution. Fresh water stands a serious risk of reduction due to the lack of recycling services offered by the environment itself. Climate change and temperature is also a very serious effect of environmental degradation. Climate change affects greatly the earth's water supply. The process of transpiration from which plants take in carbon dioxide and release oxygen has a direct relationship with the use of water. Temperatures rising can't reduce the levels of snow and also the snow seasons in the winter which increases the intensity of the melting of the snow in the Polar Regions. This in turn affects the soil moisture, a possibility of flooding and drought, as well as the storage capacities depending on the area. Warmer winters continue to increase which decreases the signal levels on mountain peaks. Lack of snow in the Polar Regions greatly affects the wildlife settled in those areas. While most of the attention will always go to climate change, global warming raises more alarm as well as the greenhouse effect. Highly populated areas continue to be at risk of environmental disasters and also pose a real threat due to the activities carried out in these places. Air pollution poses the greatest risk to the O-Zone layer. This shoots up the level of atmospheric degradation (Weil, 2004).

Overpopulation, deforestation, landfills, consumption behavior, waste pollution, lack of education, illegal dumping, agricultural pollution, littering, mining, plastic pollution and resource depletion are key factors of the constantly suffering environment leading to its degradation. There are a number of environmental disasters that we are more prone to due to the environmental degradation. Global warming, famine, wars, economic effects, loss of local and economic tourism, loss of livelihood for man, spread of diseases, poor health facilities, soil erosion, flooding, acid rains, and biodiversity loss are the major effects of environmental degradation (Srinivas, 2015).

We can recommend some solutions that can be used for the reversal and restoration process of the environment. The putting in place of some of the following solutions can help reduce the risk reduction factors and vulnerability of the human population and different areas to environmental disasters. They are:

- Entire stopping and curbing of deforestation;
- Taxes and fines for any environmental violations;
- Development of government regulations;
- Introduction of nature reserves and biotopes;
- Regulation and reduction of consumption levels of resources;
- Reduction or waste production;
- Development of recycling facilities;
- Banning of plastic selling, production, and use by man;
- General environmental education for people available on all fronts;
- Having all of humanity on the same page when it comes to environmental restoration and conservation.

Looking at the relationship between environmental disasters environmental degradation, they go hand in hand. The various factors that cause environmental degradation link us directly to environmental disasters and increase our vulnerability to these disasters. The constant exploitation of human resources by the growing number of human populations is also a major worrying factor that must be taken into consideration in the event that humanity would want to remain stable and for the earth to continue sustaining life.

In conclusion, we have gone through an array of causes leading up to the point of vulnerability and what makes us susceptible to environmental disasters. The basic points to take home would be to have the general population educated and equipped sufficiently on the care of the environment. They also need to be educated on the different disasters caused by the environment and also by the human hand and ways to deal with these disasters in the event that they strike. Equipping the human race with survival skills in the face of a hurricane, a volcanic eruption, a flood, an earthquake, an avalanche, or a landslide could be crucial to saving of lives and also prevention of these environmental disasters from happening. Generating an environmentalist ideology among the general population keeps us at bay from environmental disasters. We are careful how we consume the resources provided by the environment and we are also innovative when it comes to the economy and technology to come up with ways of recycling those renewable resources that are also available to us.

There's also an outcry for better governance by different governments four different countries all over the world. It is important for these countries and especially their governments to keep track of the environmental information and also to provide this same information for the education of the general public with regard to the environment. A provision of a strong economic system creates better communities and very strong local preparedness in the event of an environmental disaster and a quick rate of recovery for these affected areas. As for the macro factors which are population, urbanization, and environmental degradation, there is a need to keep track of energies spent on a daily basis to create revenue for different countries and also maintain and sustained life for their citizens. International environmental organizations and non-governmental organizations should create a social movement and lobby people to be oriented to save the environment which reduces the risk factor and we are less vulnerable to environmental disasters. There is a need for regulations to be put in place for factories and industries that account for majority of the pollution that is attributed to environmental disasters caused by the human hand. First world countries or developed countries are also being called upon to be careful and cautious in the rates of pollution of whatever kind.

5

Environmental Disaster Preparedness

CONTENTS

5.1. INTRODUCTION

Environmental disaster preparedness refers to taking measures for the preparation of and reduction of the effects that are brought about by disasters. This means having the ability to make predictions and if possible prevent disasters, ensure that their impact on populations that are vulnerable is mitigated, and also make sure that the consequences of these impacts are effectively coped with and responded to.

Environmental disaster preparedness helps in providing a platform for designing coordinated, realistic, and effective planning; helps in reducing effort duplication; and also helps in increasing the general effectiveness of the community members,' households' and National Societies' disaster response and preparedness efforts. It is possible to prevent disaster situations in addition to saving the most livelihoods and lives during disaster stricken situations, thus providing the affected population with a shorter time period to get back to normal (Emerald Group Publishing Limited, 2015) (Figure 5.1).

Figure 5.1: There are millions of people each year that are affected by environmental disasters on a national, local community, business, and personal level. The only way we will able to lower the impact is through successful disaster management at all levels through raising awareness, development of actions plans and practicing these plans.

Source: https://www.cdmsmith.com/en/Client-Solutions/Insights/10-Steps-to-Disaster-Preparedness.

Environmental disaster preparedness is an integrated and continuous process that results from many different resources and activities that are used for risk reduction instead of using a specific sectoral activity on its own. Many different areas are required to contribute—with a range from logistics and training, to institutional development, to livelihood, health care, and recovery.

There are many different risks that cities are exposed to when it comes to environmental exposure. There are some that are directly related to the urban settlement characteristics; there are others that many occur due to external factors, such as: extreme weather events, floods, earthquakes, or climate change. For example, if a city gets hit by a disaster such as an industrial accident, and earthquake, or a hurricane, it can greatly lead to a major negative impact on the lives and livelihood of millions of people that live in that area, in addition to having a huge effect on the economy of the country (Emerald Group Publishing Limited, 2015).

Urban sprawl and population growth can lead to spatial expansions of urban areas to areas that are prone to risks, especially those that are not suitable for human settlement, such as riverbanks, inundation zones, and slopes, all which have a high level of exposure to landslides, flooding, and erosion. In this situation, the poor in urban areas often have no option other than settling in areas that have these high risks.

So as to lower the level of risk, zoning regulations and land use planning, as well as bylaws and building codes have all been put in place in many countries and cities all over the world. However, in economies that are in transition and low income countries, the chances of having existing measures are low, and even when these measures exist, they may be outdated and/or hardly executed. The ability to assess the vulnerability and risks of local communities should be considered the first step in raising awareness of, preparing for and reacting appropriately to environmental disasters. Despite the fact that it may be difficult to avoid disasters completely, it is possible to minimize certain risks. The loss of infrastructure, homes, and lives can be reduced through risk awareness and preparedness (Cutter, 2012).

It is very important to note that a disaster could strike at any time and place in the world, all through the year, on different scales. Some the key players during the occurrence of these events are nonprofit organizations. They help by providing resources and aids to the individuals that have been affected. Indeed, animal and human shelters, food banks, hospitals, and other nonprofit programs and organizations have undergone integration into majority of the recovery and response efforts that are put into the management of emergencies.

Most of the times, we are not able to foresee these events, and this means that there is very little time, if any, to prepare for them. This is why environmental disaster preparedness is very important, and a lot of time needs to be put into preparation and planning way before there is an occurrence of a disaster (Birkmann, 2007).

5.2. REDUCTION AND MANAGEMENT OF DISASTER RISK

Disaster risk reduction (DRR) is the policy objective that anticipates and reduces risks. Despite the fact that DRR is used in an interchangeable manner with disaster risk management (DRM), the latter can generally be used to refer to the implementation of DRR, because it helps in describing the actions that should be taken in achieving the objective of risk reduction. The susceptibility to disaster indicates poor development, and therefore reducing the risk of disaster requires for the integration of policies for the reduction of Disaster Risk and practices for management of Disaster Risk into the sustainable development goals.

5.2.1. What Is Disaster Risk Reduction (DRR)?

In the past, emergency responses were the main focus when it came to dealing with disasters; however, towards the end of the 20th century, we started to notice that disasters do not occur naturally (even if the hazard in association is natural) and we are only able to alleviate the impacts of disaster and prevent losses through the management and reduction of conditions of vulnerability, exposure, and hazard (Birkmann, 2007) (Figure 5.2).

Figure 5.2: Disaster risk reduction involves the identification, assessment, and reduction of risks that are caused by environmental disasters.

Source: https://www.eventbrite.com/e/training-on-disaster-risk-reduction-tickets-98367149793.

Due to the fact that we are not able to reduce how severe a natural disaster will be, the main way in which we are able to reduce the risk is by reducing exposure and vulnerability. These two components of risk can be reduced, and it requires the identification and reduction of underlying factors that cause risk, most of which mainly in relation to urban and economic development choices and practices, environmental degradation, climate change, and inequality and poverty, all which lead to the creation and exacerbation of conditions in relation to vulnerability, exposure, and hazard. There is a need to address these underlying risk development factors so as to maintain the sustainability of development through lessening the impacts of climate change and reducing disaster risk (Veenema, 2003).

It is important to note that we need not only to manage disasters, but also risks. DRR is an important part of sustainable development and therefore it requires the involvement of every part of the private and professional sector, non-governmental organizations, the government, in the society. This means that it requires a multi-sectoral and people-centered approach which will help in creating a culture of resilience and prevention and building resilience for interacting, cascading, and multiple hazards.

In addition, DRM is inclusive of strategies that aim at:

- Avoidance of new risk construction;
- Addressing risks that is pre-existing;
- Sharing and spreading risks so as to ensure the prevention of losses from disaster from being absorbed by additional outcomes of development and the creation of a higher level of poverty.

Despite the fact that DRM is inclusive of response activities and disaster preparedness, it is all about much more than the management of disasters.

So as to achieve successful results from DRR, there's a need to combine top down, institutional strategies into changes, with bottom up community and local-based approaches. DRM programs should not be implemented on their own; however, they should be integrated within the practices and planning of development, due to the fact that disasters indicate development that is skewed or failed within the social and economic processes that are either unsustainable or from societies that are ill adapted.

The underlying risk drivers, various risk layers (from intensive to extensive risk), as well as the tailoring of local contexts, need to be addressed through all the different approaches. It is not possible to use an approach that assumes that one method will work for every group in DRM because there are many different approaches that exist, all which have undergone effective implementation for the reduction of disaster risk. However, before we are able to reduce risk, there is a need to have a good understanding of the vulnerability and exposure of assets and people to hazards and the hazards themselves (Veenema, 2003) (Figure 5.3).

Figure 5.3: Above is the disaster risk management cycle which involves responding to the disaster, rehabilitation, and recovery, prevention, and mitigation, and preparedness.

Source: http://www.un-spider.org/risks-and-disasters.

5.2.1.1. The Role of Science in Disaster Risk Reduction (DRR)

Over the past 30 years, scientific knowledge around the intensity and sharing in time or space of environmental hazards and technological ways of challenging them has grown significantly. The dramatic progresses in understanding of the main causes and limitations of environmental phenomena and methods for repelling their forces have been presented.

The United Nations for instance has proclaimed efforts towards using the existent scientific or technical knowledge, including new knowledge as required, to underpin adoption and application of public policy systems for disaster prevention. Progresses in science and technology have made it conceivable over the past few years to introduce substantial changes in integrated approaches to the subject of environmental disasters.

Science and technology allows us to comprehend the model of environmental hazards from an atmospheric, geological, biological, or hydrological origin and evaluate the change of these hazards into full disasters. Technology is also helpful in mitigating the disaster before it happens, by predicting its occurrence beforehand.

5.2.2. How Is Risk Reduced?

Disaster risk management is inclusive of activities that are in relation to:

- **Prevention:** Measures and activities that help in avoiding the new and existing disaster risks. This method always costs less than disaster response and relief. For example, the relocation of exposed assets and people far from areas that is a hazard.
- **Mitigation:** This involves limiting or lessening the adverse effects of hazards and any disasters that occur in relation to these hazards. For example, implementation of building construction codes and strict land use, planting of trees for the stabilization of slopes and construction of flood defenses.
- **Transfer:** This is a process that involves informally or formally changing the financial results of certain risks to a different party where a state authority, enterprise, community or household end up obtaining resources from the previous party after the occurrence of a disaster, in exchange for compensatory or ongoing financial or social advantages provided to the other party, for example, insurances.
- **Preparedness:** This refers to the capacities and knowledge of organizations, individuals, and communities; professional response; and governments have when it comes to effectively anticipating, responding to, and recovering from the effects of current, imminent or likely hazard conditions or events. For example, preparation of emergency supplies, identification of proper evacuation routes and installation of early warning systems.

These measures and activities are hardly implemented alone and require several different associated activities, including:

- Identifying and measuring the risk of disaster.
- Knowledge and education development.

- Raising awareness by informing people about the possible risks.
- The incorporation of DRM into national investment and planning.
- Having to strengthen legislative and institutional arrangements.
- Provision of financial protection for businesses and individuals that are at risk. This is referred to as contingency and finance planning.
- Integration of DRR across many different sectors, such as environment, heath, finance, etc.

We are able to describe the activities that are used for risk reduction as either: structural, such as implementation of building codes and land use planning; and non-structural, such as legislation, policy making, and raising awareness. Risk governance refers to the manner in which the civil society, governments, and other players in this sector organize for the management of disaster risk, for example through mechanisms for accountability and participation, decentralization, and legislation, and institutional arrangements. Research has been able to provide clear evidence that suggests that weak governance in low income countries makes these countries less resilient and more vulnerable to disaster risk (Cutter, 2012).

More importantly, DRR helps in risk reduction by ensuring the growth of the resources, attributes, and strengths that are found within an organization, society, or community-which is generally referred to as their capacity. Activities for the management of disaster risk have been created in a manner that allows them to help in increasing the resilience of systems, society, communities, and people to recover from, accommodate, absorb resist, and improve wellbeing especially in this time with many hazards.

Activities for the reduction and management of risks can help in providing means for improving resilience to other common risks. Apart from development, it is also important to ensure that DRM is also integrated across many different sectors, inclusive of conflict and climate change (Coppola, 2010).

5.3. IDENTIFICATION AND UNDERSTANDING OF RISK: THE MOST IMPORTANT ASPECT OF REDUCING RISK

Understanding, Identification, awareness, and measuring risks that may occur due to disasters are all very important aspects of DRM. DRR is all inclusive of choosing and making decisions, as well as lack of choices and decisions (Figure 5.4). This means that risk information is very important in five major areas of making decisions. These include:

- **Identification of Risk:** Due to the fact that the losses and damages that are due to historical disasters are not commonly well identified, and because the potential loss and damages that could occur from disasters that may happen in the future (such as high-impact events that are not frequent) may not be known at all, DRM is ranked at a low level of priority. Appropriately communicating robust risk information at the correct time can help in raising awareness and also triggering action.
- **Risk Reduction:** Risk and hazard information can be used for informing many different activities for the reduction of risk, from designing risk reduction measures such as storm and flood surge protection and improvement of building codes, to carrying out assessment risk different building types at the macro level, for example when it comes to prioritization of investments in retrofitting and reconstruction.
- **Preparedness:** The comprehension of the affected geographic area, together with the frequency and intensity of various hazard events is very important when it comes to running preparedness drills, creating shelters, and planning evacuation routes. Being able to provide a measure of the negative effects that have been caused by different hazard events—secondary hazards, injuries, and fatalities, potential number of buildings that are damaged—provides the ability for the establishment of plans that are realistic and detailed for better disaster response systems, which can generally help in reducing the severity of environmental disasters.
- **Financial Protection:** Insurance and financial sectors helped in the development of disaster risk analysis especially when it came to quantifying risk of high impact natural hazard events that are comparatively rare. All this is important because of the high need for governments to ensure management of their support programs or sovereign financial risk that help in managing individual financial risks, for example household earthquake insurance or micro-insurance.
- **Resilient Reconstruction:** Assessment of risks can be very important when it comes to impact modeling before disasters

occur, for example in the days that lead up to a hurricane, or it can also help in providing rapid and initial estimates of economic, physical, and human loss in the immediate aftermath of an event. In addition, there is a need for risk information before an event occurs so as to achieve resilient reconstruction, because there is hardly any time after an event for the collection of the required information for informing land-use plans and resilient design (Figure 5.5).

Figure 5.4: Before you understand how to be prepared for environmental disasters, it is necessary to comprehend what disaster risk is.

Source: https://www.preventionweb.net/risk/environmental-degradation.

Figure 5.5: Resilient reconstruction is important in reducing and preventing the loss of life and assets. It is important for engineers to consider impact modeling especially in high risk areas.

Source: https://www.worldbank.org/en/events/2016/04/26/resilient-reconstruction-in-nepal-transforming-risks-to-resilience.

If the individuals that are being exposed to these hazards are not aware of the risks that they may face, it is not easy to see why or how governments, businesses, or households would make investments in reducing the levels of risk. However, despite the fact that awareness of these risks may be a precondition, the level of importance that these individuals attach to the management of the risks can only be comprehended in the full range context of the environmental, territorial, economic, and social opportunities and constraints that they face.

There is generally around 30 years of research that has been done in disaster risk; however, most of this information is not available in a form or manner that can be useful or understandable to the individuals who require this information the most. This means that it is important for risk researchers and scientists to pay more attention to producing risk information that is actionable and understandable by different users in terms of words and risk knowledge. For this shift to occur there is a need for more partnership and collaboration between researchers and scientists as well as those who are involved in DRR, ranging from local communities to governments (Weil, 2004).

There is a need for governments to invest in collecting, managing, and disseminating risk information, including impact and disaster loss statistics, vulnerability information, exposure databases, and hazard models. In addition, governments need to put in place mechanisms and standards that help in showing transparency and openness so that the individuals are not only able to access the information that they require, but are also well informed on the underlying limitations and assumptions. The actionable and understandable risk information needs to be generated in a particularly sensitive manner when it comes to extensive risk, which, due to the fact that it is configured to an extent that is much larger by environmental, economic, and social vulnerability, can be effectively lowered through sustainable development practices and risk management.

As stated above, there is a need to understand the factors involved in risk management and reduction. These will be explained below. But first, what is disaster risk?

5.3.1. Disaster Risk

This is expressed as the possibility of damage and destruction, injury or loss of life from a disaster in a certain time period. It is mainly explained as the result of a hazard and the elements that make places and people exposed and vulnerable. There are times in which disasters are considered external shocks; however, disaster risks are due to interactions that are complex, between processes for development that help in generating conditions of hazard, vulnerability, and exposure (Weil, 2004).

This means that disaster risk refers to the combination of the frequency and severity of a hazard, the number of assets and people that have exposure to the hazard and how vulnerable they are to the damage. Disaster risks that are associated with high impact events and have a low probability are referred to as intensive risks. Those associated with low impact events and have a high probability are referred to as extensive risks. The impacts and losses that characterize disasters normally have a lot to do with the vulnerability and exposure of places and people as they do with how severe the hazard event is (Cutter, 2012) (Figure 5.6).

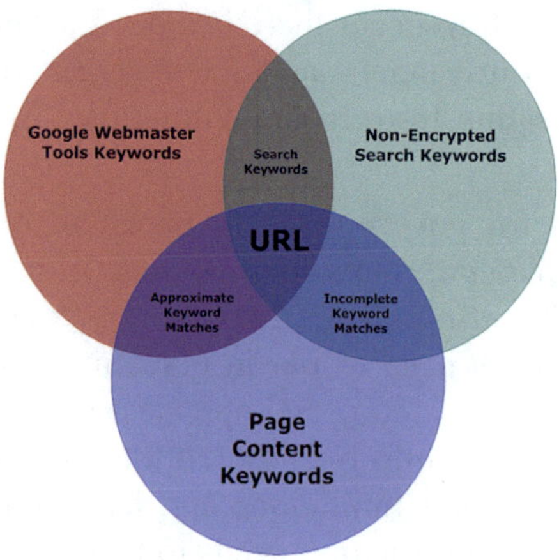

Figure 5.6: The three components of disaster risk are exposure, vulnerability, and hazard. The three have to be available for a certain aspect to be considered a risk.

Source: http://www.oilandgasbmps.org/resources/PH_Risks.php.

There are several characteristics of disaster risk. So as to fully understand disaster risk, these need to be well understood. They include:

- Forward looking at the possibility of damage and loss of life in a given time period and loss of life.
- **Dynamic:** It can decrease or increase in accordance with the ability to lower vulnerability.

- **Invisible:** Not only does it comprise of the high impact event threat, but also the low impact events that frequently occur and are often hidden.
- **Uneven Distribution Around the Earth:** Different areas are affected by hazards; however, the disaster risk pattern reflects the exposure's and vulnerability's social construction in various countries.
- **Complex and Emergent:** Many different processes, including globalized economic development and climate change, are leading to the creation of new, interconnected risks.

The ability to identify, assesses, and understands disaster risk is important in its reduction. The disaster risk can be measured through making analysis of trends of, for example, previous losses due to disasters. The trends can allow for us to see whether the system of DRR is actually working. Future losses can also be estimated through risk assessment conduction.

A risk assessment that is comprehensive works effectively when there is consideration of the whole range of disaster events that can potentially occur as well as their underlying uncertainties and drivers. This can all start by analyzing the historical events, the incorporation of forward-looking perspectives, as well as the integration of anticipated impacts of phenomena that have led to the alteration of historical trends, including climate change (Srinivas, 2015).

Additionally, risk assessment should include the consideration of events that are rare—those that are positioned outside the future projections of hazards—but that could still happen based on scientific knowledge. Anticipation of events that are rare needs a wide range of interdisciplinary findings and information, together with simulations and scenario building, which can all undergo supplementation by expertise from many different disciplines.

Information on losses, vulnerabilities, exposures, and hazards help in enhancing the accuracy level of risk assessment, thus leading to the contribution of measures that are much more effective for the prevention, preparation for and financial management of disaster risk. The modern approaches that are commonly used in risk assessment include risk modeling which was brought about by the development of computational resources in terms of power and availability. These models provide us with the ability to create simulations of the likelihood and outcomes of various events (Birkmann, 2007).

The production of risk assessments is necessary so as to make estimations of the possible social, infrastructure, and economic impacts that are coming up from one specific hazard or many other hazards combined. The three components of risk assessment as well as the associated losses are hazard, exposure, and vulnerability. These are explained in subsections (Figure 5.7).

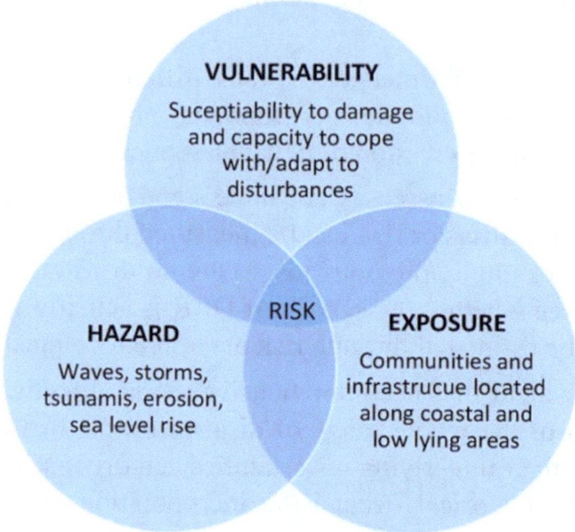

Figure 5.7: Above is a good example of the interaction between hazard, vulnerability, and exposure when it comes to disaster risk.

Source: https://www.researchgate.net/figure/The-interaction-of-three-factors-vulnerability-exposure-and-hazard-that-influence_fig2_320674236/amp.

5.3.1.1. Hazard

This is referred to as a human activity, phenomenon, or process that may lead to environmental degradation, economic, and social disruption, property damage, injury or other health impacts, or loss of life. Hazards may be socio-natural, anthropogenic or natural in origin.

Hazards mainly undergo categorization by whether they are technological (also referred to as human-induced or man-made) or natural (also referred to as physical). Instead of hazard, there are times when the word 'peril' is used, especially in the insurance industry.

Physical (or natural) events are only referred to as hazards when they have the ability to cause economic and social disruption, property damage, or harm to people. The area in which a natural hazard occurs is mainly dependent on processes that are natural, including the existence of slopes and waterways (which may lead to landslides), the influence of weather systems, and the movement of tectonic plates. However, some processes such as climate change, environmental degradation, and urbanization can also lead to an influence in the intensity, frequency (occurrence) and location of natural hazards. These are referred to as risk drivers (Birkmann, 2007).

The schemes that are used for the classification of hazards are different in various governments and research institutions; however, these can be divided into:

- **Biological Hazards:** These are those that are conveyed by biological vectors or of organic origin, including bioactive substances, toxins, and pathogenic microorganisms. Some of these include parasites, viruses or bacteria, as well as mosquitoes, poisonous plants and insects that carry agents that cause diseases, and venomous wildlife.
- **Environmental Hazards:** These are inclusive of biological, natural, and chemical hazards. These can either be formed due to chemical or physical air, soil, and water pollution, or environmental degradation. However, multiple of these phenomena and processes that are in this classification may be referred to as drivers of risk and hazard instead of hazards themselves, for example sea-level rise, salinization, loss of biodiversity, deforestation, and soil degradation.
- **Geophysical or Geological Hazards:** These mainly come from processes within the earth. For example, emissions, volcanic activity and earthquakes, and other processes such as mud flows, debris, and surface collapses, rockslides, landslides, and mass movements.
- **Hydro-Meteorological Factors:** These are major contributors to some of the previously mentioned processes. It is difficult to categorize tsunamis: despite the fact that they are triggered by earthquakes under the sea and other geological events, they end up being a process in the ocean that is then considered a coastal water related hazard. These hazards are of oceanographic, hydrological, or atmospheric origin. For example, tropical cyclones that are also referred to as hurricanes or typhoons; coastal storm surges; heat waves and cold spells; droughts; and floods, including

flash floods. These conditions can also lead to the causation of other factors such as epidemics, locust plagues, wild-land fires, landslides, and also in the dispersal and transport of substances that are toxic and volcano eruption material (Figure 5.8).

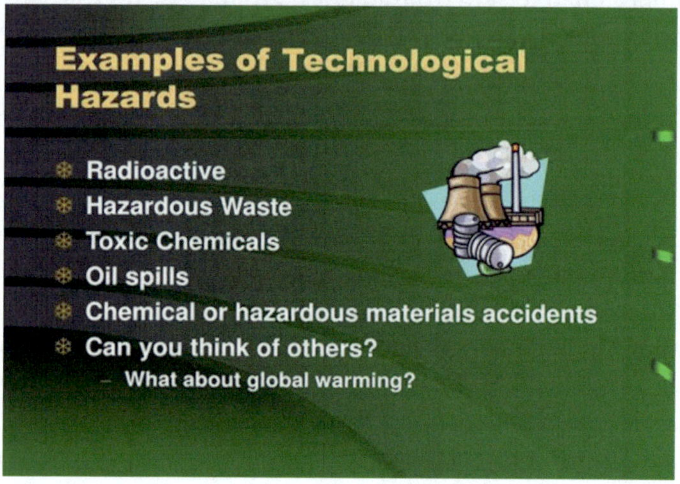

Figure 5.8: Above are some common examples of technological hazards. These can cause different forms of pollution leading to climate change that can lead to several environmental disasters.

Source: https://www.slideserve.com/hedva/root-causes-of-social-vulnerability-technological-and-human-induced-hazards.

- **Technological hazards:** These stem from industrial or technological conditions, specific human activities, infrastructure failures, and dangerous procedures. Some of the examples include chemical spills, fires, factory explosions, transport accidents, dam failures, toxic wastes, nuclear radiation, and industrial pollution. These hazards may also occur in a direct manner due to negative impacts from natural hazard events.

Generally, one hazard can end up triggering several other hazards, for example tropical cyclones (which are also referred to as typhoons in the Northern Pacific Ocean, cyclones in the Indian Ocean, and hurricanes in the Atlantic Ocean) can end up causing heavy rainfall, storm surge and intense winds, in addition to triggering secondary hazards such as landslides. A cascading or domino effect can be caused by a series of relationships that are triggering, for example, the 2011 Japan crisis that was later referred to as the tsunami-earthquake-nuclear crisis (Veenema, 2003).

We can characterize natural hazard events by their intensity or magnitude, the area they cover, duration, and speed of onset.

There is different magnitude or intensities at which hazards occur, over different temporal scales or time scales. Scientists use return periods or probabilities to refer to the different intensities at which hazards occur within the context of uncertainty. These are also referred to as recurrence intervals. Generally, the less frequent the hazard (the longer the period of return), the larger the hazard intensity. Due to the long periods of return, some communities do not remember how potentially harmful this high intensity hazard may be.

There are also different geographical scales at which hazards may occur. For example, tornadoes tend to occur in areas that are quite localized, and droughts tend to occur for more than a couple tens of thousands of kilometers. There are many different countries that have a high exposure to many different hazards. This means that making considerations of the specific risks that are related to the full hazard range that may affect assets or people is important. Sadly, there have been several situations in which there has not been any consideration to the full range of hazards; for example, after the Indian Ocean tsunami that occurred in 2003, there was still house reconstruction in areas that are prone to floods in Indonesia, which meant that the families living in these houses were later exposed to the other tsunami that occurred in 2007 (Veenema, 2003).

There is also a possibility for hazard interaction; for example, in the Philippines, when Mt. Pinatubo erupted in 1991, this event was accompanied by a typhoon in which the rainfall ended up soaking up the cumulative volcanic ash-fall. This soaking up lead to an increase in the weight of the falling rainfall which then caused the roofs of businesses and homes to collapse, which resulted in approximately 300 deaths, all in direct association with Mt. Pinatubo erupting. This shows that hazard interaction can cause a much greater negative effect in comparison to the two different hazards occurring at separate times. This aspect has huge implications when it comes to risk assessment.

Location and, in the situation of some hazards, anticipation over the different time periods in which some hazards may occur can be achieved through the improvement of hazard knowledge, as well as the conduction of hazard assessments. The ability to anticipate ranges from probability analysis of hazard occurrences that are long term, to the hourly, daily or monthly monitoring a d detection of hazards, so as to effectively inform the early warning systems (Paul, 2011).

It is important to ensure that warning systems are accompanied by strategies that have been put in place for DRR so as to ensure the reduction of vulnerability and also the enhancement of the capacity that people have to respond to and recover from a disaster. If a slow-onset hazard was to occur, it is possible to use early indicators of the potentiality of a crisis for the detection of the warning signs, so as to help in building resilience (Figure 5.9).

There are four components if early warning systems. These include:

- Detecting, monitoring, and providing a forecast of the hazards.
- Analyzing the risks that are involved.
- Disseminating warnings that is both authoritative and timely.
- Activating the emergency preparedness and response plans.

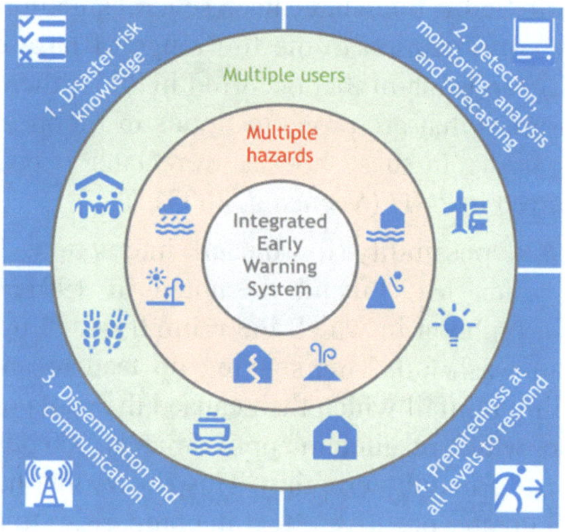

Figure 5.9: There are four main characteristics of effective integrated early warning systems. These include: disaster risk knowledge; detection, monitoring, analysis, and forecasting; dissemination and communication; and preparedness at all levels to respond.

Source: https://public.wmo.int/en/resources/bulletin/role-of-young-professionals-driving-integration-of-early-warning-systems.

There are many different agencies that are needed for the effective coordination of these components and also for them to work; both at the community and national levels. If one component fails, or if they are not well coordinated, the whole plan can fail.

5.3.1.2. Exposure

These refers to the situation of production capacities, housing, infrastructure, people, and other human assets that are tangible found in areas that are prone to hazards. If there is occurrence of a hazard in an area that is not exposed, then there will be no risk. Generally, the determination of the extent at which people who are exposed or economic assets are actually at risk is all dependent on their vulnerability. This is because there is a possibility for exposure but not vulnerability (Cutter, 2012).

However, there has been some evidence suggesting that in the situation of an extreme hazard, the disaster risk degree is due to exposure more than it is due to vulnerability. For example, the Indian Ocean tsunami that occurred on 26th December 2004, affected all the exposed individuals, without paying any attention to the social class, ethnicity or income.

Economic assets and people normally increase in concentration in hazard exposed areas through processes such as economic development, urbanization, migration, and population growth. Exposure can also be driven by previous disasters because they may force people put of their lands to areas that are very unsafe. At the same time, exposure changes from place to place and over time.

Many areas that are prone to hazards such as flood plains, volcanic slopes and coastlines, lead to the attraction of urban and economic development, are of religious or cultural significance to individuals living in that area, or offer significant economic benefits. With increased exposure to people and assets, comes a much more concentrated risk in these areas. At the same time, the expansion of cities and urban and economic development increases the spread of risks due to the transformation of areas that were previously sparsely populated (Cutter, 2012).

There is still a high volume of capital that is flowing into areas that are prone to hazards, which significantly increases the value of economic assets that are exposed. If there is a continuous upwards trend in global exposure, there may be an increase in the disaster risk to a level that is very dangerous (Figure 5.10).

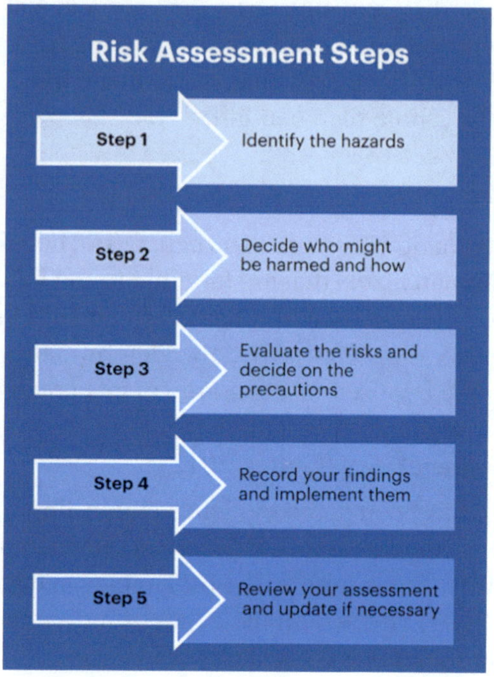

Figure 5.10: Above are the five main steps in achieving proper risk assessment. Risk assessment is important in identifying the hazard, level of exposure, and the level of vulnerability, making it a very important step of creating disaster preparedness systems.

Source: https://www.lucidchart.com/blog/risk-assessment-process.

We can measure exposure either through the types of assets that are found in the area, or the number of people that are found in the area. We can then combine these with the capacity and the specific vulnerability of the elements that are exposed to a specific hazard so as to make an estimation of the quantitative risk that is in association with the hazard that may occur in the specific area (Birkmann, 2007).

When it comes to risk assessment, exposure modeling has proven to be very useful. The data that is utilized for the development of exposure data sets can be gotten from many different methods and sources. The data set resolution helps in referring to how well define the data is; for example, the true picture at the local level may be hidden by a low resolution (exposure data that is over a wide geographical area).

Locally, there has been development of high resolution exposure data; however, this can only be used for specific projects. Some of the commonly used sources of data include individual structural/architectural drawings, household surveys, and local and council government agencies.

The process of using the Internet to obtain data or input into a task by hiring these services from a many different individuals, also referred to as crowd sourcing, has become a valuable and common tool in the collection of data in the local level, and also in the validation of data on a global scale; however, the method is limited by the data quality and type (Birkmann, 2007).

There is an upward trend in the economic exposure that is occurring in high-hazard areas. If this trend is not reversed, there will be an increase in the disaster risk. It is important for us as the inhabitants to act now to ensure the reduction of this exposure and also to purpose on building resilience and capacity in these areas that have a growing exposure.

When there is no ability to prevent exposure to disasters, location decisions, and land use planning needs to be accompanied by other non-structural or structural methods to mitigate or prevent risk. For example, in the Indian Ocean tsunami in 2004, reducing exposure through timely evacuation would have been the only way to save lives, and this was dependent on the availability of effective and reliable preparedness planning and early warning systems, and thereafter loss compensation through insurance of other instruments for financing risks (United Nations, 2010).

5.3.1.3. Vulnerability

The characteristics of vulnerability are determined by environmental, economic, social, and physical processes or factors that cause an increase in the level of susceptibility that a system, asset, community or an individual has to the impact of hazards. It is important to understand what makes people vulnerable (Weil, 2004).

Vulnerability can be defined in this situation as the human dimension of disasters and is due to the range of psychological, political, institutional, cultural, social, and economic factors that shape the lives of people as well as the environment in which they live.

This concept can be challenging to understand mainly due to the fact that it tends to have various changes to its definition with the different people and communities, and because of this it may hand many different terms to describe it, including 'lack of capacity,' 'deficiency,' 'weakness,' 'fragility,' and 'predisposition' (Figure 5.11).

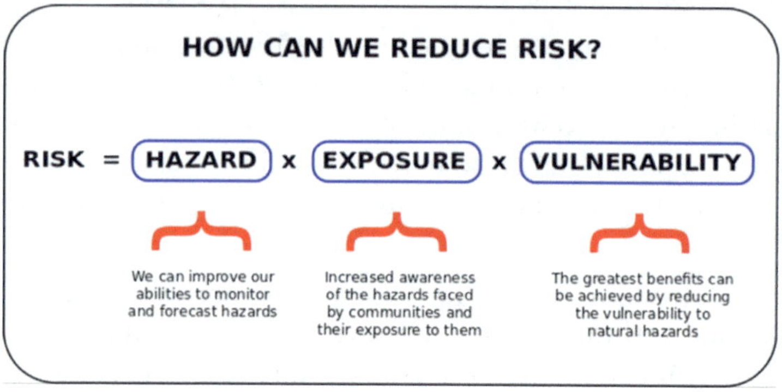

Figure 5.11: The image above provides a visual explanation of how we can reduce risk. When it comes to the aspect of vulnerability, it is considered the greatest aspect in the reduction of risk to natural hazards.

Source: https://www.researchgate.net/figure/Disaster-Risk-Hazard-exposure-and-vulnerability-are-all-important-parts-of-the-risk_fig4_313185393.

In addition to susceptibility to harm, some definitions have been able to include exposure. It has however been recently understood that the susceptibility element of vulnerability is separate from exposure because it possible to have exposure and at the same time lack susceptibility to the natural hazard.

Regardless of the differences over the definition of vulnerability, many experts in this field have been able to conclude on the fact that the comprehension of vulnerability needs more than just analyzing the hazard's direct impacts. The wider social and environmental conditions that limit communities and people when it comes to coping with the impact hazard are also concerned with vulnerability (Weil, 2004).

It is important to understand that poverty is not the main aspect of vulnerability, however, research has been about to prove that in majority of the cases, the poor normally suffer from disasters the most. Poverty can be considered both a consequence and driver of disaster risk (especially in those countries that have risk governance systems that are weak) due to the fact that pressure from the economy pushes people to live in areas and conditions that are unsafe. The different multi-dimensional factors, including poverty, and drivers that lead to the creation of vulnerability meaning that most of the time, but not always, there is an association between the susceptibility to the impacts of hazard and certain groups including displaced populations, migrants, the disabled, the elderly, children, and women, amongst others.

There are a number of factors that are related to vulnerability, including:

- **Physical Factors:** These include unregulated land use planning, poor design and construction of buildings, etc.
- **Social Factors:** These include psychological factors, age, and disability, social status, discrimination by gender and social exclusion, marginalization, and inequality and poverty, etc.
- **Economic Factors:** These include globalization of supply and business chains, dependence on single industries, vulnerable rural livelihoods, the uninsured informal sector, etc.
- **Environmental Factors:** These include climate change, decline of risk regulating ecosystem services, overconsumption of natural resources, poor environmental management, etc.

In addition, we determine vulnerability through natural resource and institutional, cultural, political, and historical processes that help in shaping the environmental and social conditions the people exist in. A number of immediate unsafe conditions are produced by these processes, such as a lack of preparedness measures or local institutions, political tensions, ill-health or poor housing or living in dangerous locations.

There is an increase in the underlying drivers of vulnerability, including the poor management of development in urban areas, thus leading to an increase in vulnerability in many regions and countries globally. Despite the fact that evidence has suggested they countries that are wealthier and well governed have the ability to lower disaster risks, there are some countries that have proven to exhibit a rapid economic growth without any visible reduction in the rate of vulnerability (Cutter, 2012).

The inclusion of vulnerability into the comprehension of disaster risk shows the acknowledgment of the fact that not only does disaster risk depend on the hazard's severity, or the quantity of assets or people that are exposed, but that it also works as a reflection of the fact that economic assets and people are susceptible to suffering damage and loss. The exposure and vulnerability levels help in explaining why extreme impacts and disasters can be brought about by non-extreme hazards, while some events that are extreme do not.

When it particularly comes to extensive risk, the vulnerability that people have ends up being the main factor in the determination of their risk.

When it comes to the different hazards, there are some groups that have a higher level of susceptibility to suffering, loss, and damage in comparison to others, and in the same manner (in these same groups) some specific individuals tend to experience higher vulnerability levels than others. Groups that are vulnerable have a harder time reconstructing their livelihoods after a disaster, thus making their level of vulnerability much higher to the possible disasters. Generally, for us to reduce disaster risk, we have to lower the level of vulnerability (Cutter, 2012) (Figure 5.12).

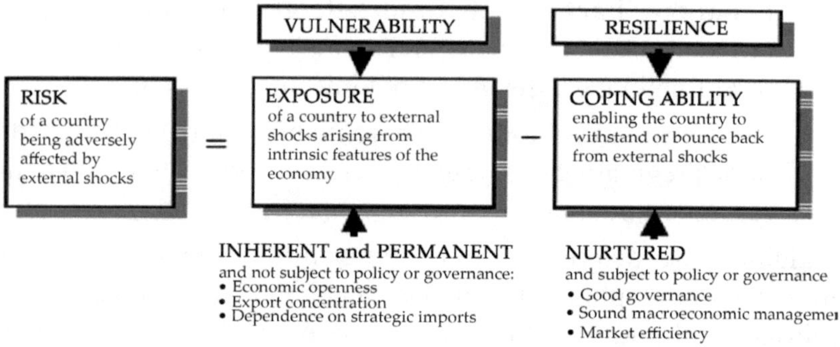

Figure 5.12: Above is a structure showing the concepts of economic vulnerability and resilience.

Source: https://www.semanticscholar.org/paper/ECONOMIC-VULNERABILITY-AND-RESILIENCE-CONCEPTS-AND-Cordina-Farrugia/d34c696836257a54f3ee8cd783bd0b8cd2e7420c/figure/5.

There are many dimensions within vulnerability, making it a very complex aspect. There are many different factors at different levels that drive it, from global to local. In addition to this, it is dynamic, due to its ability to change under the pressure of the forces that drive it. In addition, these factor the cause vulnerability amongst prone are not always obvious.

The chain that leads to vulnerability, from the immediate presenting conditions such as poor quality housing, to the underlying drivers such as socioeconomic processes, can both be complex and long; however, we are able to identify the vulnerability progression that causes community pressures by tracking it. We are then able to release these pressures by putting measures in place that help in reducing vulnerability at several points along the chain that causes it (Dufresne, 2019).

Due to the various facets, there is more than one method that can be used to assess vulnerability. Ideally, all forms of assessment that is used to asses vulnerability need to adopt an approach that is holistic. In reality, we divide these methods into those that consider socioeconomic vulnerability, and those that consider physical (or environments that are built) vulnerability.

Making assessments of the vulnerability that built environments have to hazards is paramount in predicting the potential effects that could be caused by an event, and also for mainstreaming DRR into the local process of development planning. Being able to understand how existing structures respond to potential hazards, such as wind from tropical cyclones and ground shaking from earthquakes, needs the knowledge of engineering practices and building materials. It is only possible to develop this information base in a sustainable and reliable manner at the local level.

There are many local engineers that have chosen to pay much more attention to the comprehension of the vulnerability of the building materials they are using locally (and these vary significantly within countries and from one country to another) to the various natural hazards. For example, engineers in Indonesia and the Philippines have been able to develop vulnerability calculations that are relevant to the specific building materials that they use. However, this is only representative to the exception. Sadly, there is hardly any opportunity for loss and damage data collection after the occurrence of disaster events, yet this is very important in the comprehension of future risks (Dufresne, 2019).

The efforts to quantify poverty and socioeconomic vulnerability have remained limited, and there is rarely any integration of such information into the assessment of risks. The quantification of social vulnerability has remained a big problem, however, indices, and indicators that are being used for the measurement of vulnerability have been developed (descriptive and quantified), all ranging from those used at the community level, to global indicators. They are many used for tracking any changes that occur in vulnerability over a certain time period (Babar, 2007).

5.4. DISASTER PREPAREDNESS TOOLS

5.4.1. International Federation Approach

There is a need for National Societies that are cooperating with the International Federation of Red Cross to prioritize disaster preparedness and ensure that it is integrated into the overall efforts that are put into programming. Implementing measures for disaster preparedness that are effective can contribute towards the building of resilient, safe, and well prepared communities and households, which will help in the reduction of future and existing vulnerabilities and risks (Figure 5.13).

Figure 5.13: Then International Federation has two main aims. These include: the improvement of international, regional, and local capacity to respond to public health and disaster emergencies, and ensuring a scale up in the action being taking to promote health, prevent diseases and reduce disaster risk in vulnerable communities.

Source: https://www.ifrc.org/en/who-we-are/the-movement/ifrc/.

International federations together with the partnerships that they have developed in the government have the role of improving the livelihood and lives of people who are vulnerable and are considered to be at a higher risk from events that may end up threatening their survival, or even the capacity they have to live with a socioeconomic security level that is acceptable, and human dignity (United Nations, 2010).

5.4.2. Community Preparedness

Disaster preparedness tools should be able to provide a platform for designing coordinated, realistic, and effective planning, all which will help in reducing effort duplication and lead to an increase in the general level of effectiveness of response efforts and disaster preparedness when it comes to the community members, households, and National societies.

When we embed disaster preparedness activities with measures for reducing risks, we are then able to prevent situations that are brought about by disasters, which results in saving livelihoods and maximum lives in case of a disaster, therefore making it possible for populations that have been affected to rebuild their lives in a short time period.

5.4.3. Early Warning Systems

When it comes to saving lives, early warning systems have some limitations, especially if they are not combined with networks that are people-centered. So as to work effectively, it is important for early warning systems to be relevant to; trusted by, and understood by the individuals they are serving. Unless these warning reach the people who are most vulnerable, then they have little to no value. These people need to also be trained on how to effectively respond to hazards that may be approaching. There is a need to develop efficient and effective warning systems, and International Federations should look into this. There is however a need to understand the importance of:

- Establishment of local networks that are able to act on and receive warnings as well as have the ability to educate communities on action taking and raise awareness on the possible risks, so as to make sure that they remain safe.
- Use these local networks to ensure the development of warning systems in a progressive manner, so as to make sure that the needs of the communities are met, in addition to performing other tasks that they are designed to perform.
- The ability to take a multi-hazard approach towards making sure that there is sustainability through providing relevance, awareness, and relevance (Figure 5.14).

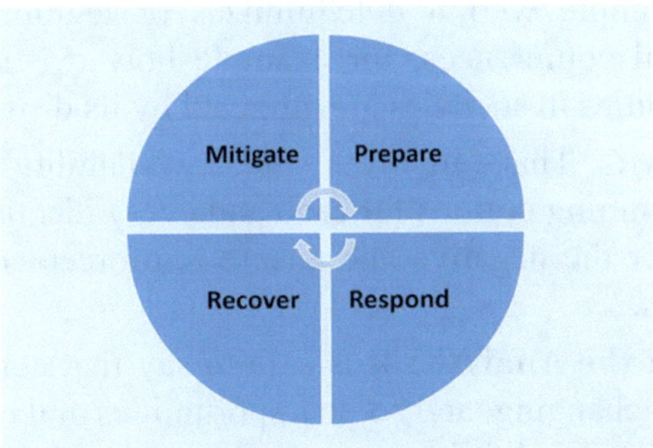

Figure 5.14: Community preparedness refers to the ability that communities have when it comes to preparation for, withstanding, and recovering from di-

sasters in both the long and the short term. Determining the risk is the main step in achieving this.

Source: http://www.masslocalinstitute.info/CommunityPreparednessAwareness/CommunityPreparednessAwareness_print.html.

5.4.4. Better Programming Initiative

This is an initiative that acts as a tool that is generally used for assessing impacts. The International Federation developed the initiative, and its adoption is from the Local Capacities for Peace Project. The Better Programming Initiative was mainly developed based on the fact that communities that were affected by violence could be supported through the provision of local capacities for reconciliation and recovery with the help of well-planned humanitarian aid (James and Paton, 2015).

The Better Programming Initiative methodology normally works through the application of five main steps:

- **Context Analysis:** This involves the identification and prioritization of the dividers (division) and connectors (cohesion) that are used to characterize the context.
- **Aid Program Description:** This involves describing planned actions in detail, such as: when, what, where, why, by whom, with whom, and most importantly, how is this aid being provided?
- **Impact Identification:** This involves considering whether the aid will weaken or reinforce the connectors and dividers. Aids normally make an impact through both its symbolic consequences, for example who it delegitimizes or legitimizes, and for its material conferences, for example how existing divisions and inequalities in societies are attracted by its distribution.
- **Options:** This involves the availability of brainstorm programming options for each and every identified impact, so as to lower the negative and ensure reinforcement of the positive impacts.
- **Repeat the Analysis:** It is safe to say that contexts are always rapidly changing, and so do opportunities and constraints for aid programming. It is important to ensure that the analyses that have been done are updated as frequently as the permits provided for the project cycles.

5.4.5. Vulnerability and Capacity Assessment

This normally utilizes many different participatory tools to ensure that the people's capacity to and exposure to resist natural hazards is efficiently gauged. It is considered a very important part of disaster preparedness and provides a huge contribution towards the creation of disaster preparedness programs that are based in the community at both the urban and rural grass-roots level.

Vulnerability and capacity assessment allows for the identification of local priorities and ensures that the appropriate actions are taken in ensuring the reduction of disaster risk while assisting in the development and designing of programs that are responsive to the people's needs in addition to being mutually supportive (James and Paton, 2015).

The vulnerability and capacity assessment aims to:

- Makes sure that hazards and risks that communities together with the capacities these communities have to deal with them are assessed.
- Involving of development organizations, humanitarians, and local authorities, and communities in making assessments from the outset.
- Drawing up of action plans that can be uses for the preparation of and respondence to the risks that have been identified.
- Identification of activities that can be used to reduce risks so as to lessen or prevent the effects of the expected vulnerabilities, risks, and hazards.

Vulnerability and capacity assessment is harmonious with sub-national and national capacity, vulnerability, hazard, and risk mapping exercises that help in the identification of communities that are most vulnerable. A vulnerability and capacity assessment is then done in the specified communities so as to find out the specific locations that are vulnerable or at risk, so as to be able to determine the actions that can be taken so as to deal with these issues. This circle is then completed when the unearthed information at the local level is then provided to the sub-national and national levels so as to provide a valuable indication of the capacities and vulnerabilities at these other levels (Figure 5.15).

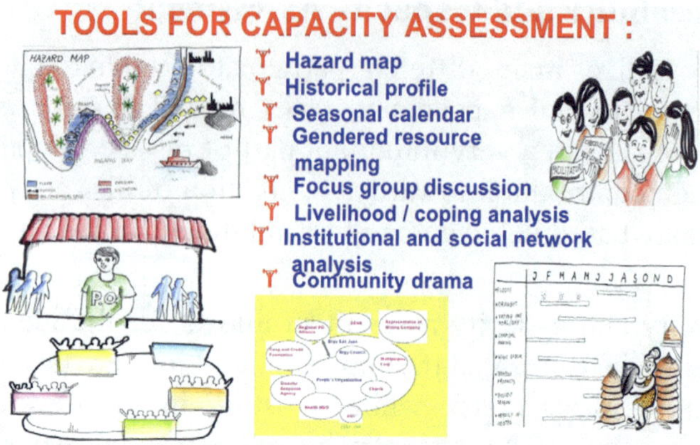

Figure 5.15: Above are some of the important tools in capacity assessment. This analysis helps in making a determination on what individuals do when disasters occur, so as to help in reducing the hazard's damaging effects, and also for the protection of lives, livelihoods, and assets.

Source: https://www.slideshare.net/NDRCNepal/hazard-vulnerability-capacity-assesstment-b.

Over the last ten years, the International Federation has been able to gain experience that has enabled it to improve and refine the Vulnerability and Capacity Assessment, so at to make sure that it better focuses on the main issues, and is able to much more effectively achieve its purpose. It has also been able to provide proof that the vulnerability and capacity assessment can be linked to and used for the reinforcement of other activities and programs, so as to improve the capacity level. In addition, due to the fact that the vulnerability and capacity assessment is a process that is participatory, it is possible for Nation Societies to develop relevant and realistic activities that are much more suited to local priorities and needs.

5.4.6. Well-Prepared National Societies

The characteristics of a well-prepared National society were defined by staff from different National Societies and disaster preparedness delegates from the International Federation in the year 2001. They mainly aimed at preparing a simple tool that can be utilized for disaster preparedness planning.

Based on this, it is safe to say that Well-prepared National Societies is a self-assessment tools that is able to provide National societies with the ability to prioritize, recognize, and plan for activities that can promote the improvement of their capacity for managing disasters (Coppola, 2010).

The Well-prepared National Societies is a project that is shared between the International Committee of the Red Cross and the International Federation. It is used to provide a better form of coordination for the preparedness support that the National Societies are provided with when they need to deal with both conflict situations and natural disasters.

The societies can be utilized for the prioritization of activities as well as for the development of indicators that can be used for progress monitoring. This is considered a precursor to tools that are much more comprehensive, such as VCA, all which can be utilized for risk reduction activities and community preparedness, as well as the mapping of capacities, risks, and hazards at the National Society level (Emerald Group Publishing Limited, 2015).

The well-prepared National Society aims to:

- Provide the National Societies with help to make assessments of their preparedness.
- Make identifications of the weaknesses and strengths, as well as ensure that actions are being effectively prioritized.
- Ensure provision of indicators and a baseline which are global and/or regional.
- Ensure identification of key areas where there is a major need for disaster preparedness support at a national, regional, and global level, so as to have the ability to include emergency and annual appeals.
- Make sure to provide negotiation inputs with donors, partner National Societies or even other partners.
- Work to make identifications of regions and national societies that have specific resources and expertise that other regions can benefit from as well.
- Identify long-, medium-, and short-term trends that can be used in disaster preparedness.
- Enabling National Societies to make assessments of their statuses against regional and/or global trends (Figure 5.16).

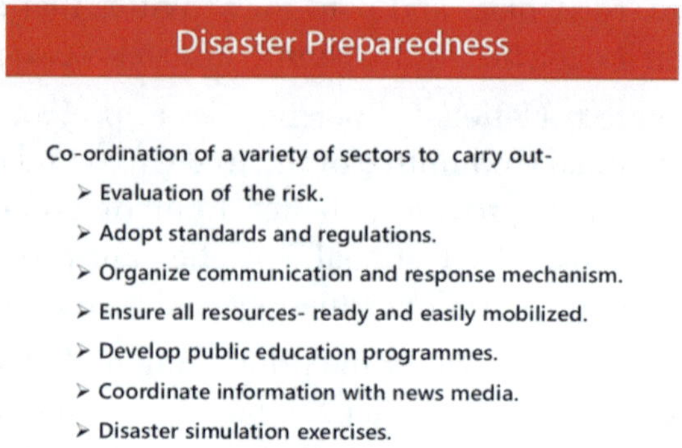

Figure 5.16: There is a need for partnership and coordination so as to achieve disaster preparedness. Some of the main functions that need to be carried out include: risk evaluation, adoption of regulations and standards, and organization of response and communication mechanisms.

Source: https://www.slideshare.net/drbalanshaikh/disaster-management-70789125.

5.4.7. Contingency Planning

This aims at preparing an organization that can be used to properly respond to emergencies as well as its potential humanitarian impact. The development of a contingency plan is all about the creation of decisions in advance, all which are on managing financial and human resources, coordinating, and communicating procedures, as well as being knowledgeable on the range of logistical and technical responses.

This form of planning is considered a management tool that involves all the sectors, and it can help in ensuring effective and timely provision of humanitarian aid to individuals who need it the most during the occurrence of a disaster. When we spend time on contingency planning, we are then able to save time during the occurrence of a disaster. Contingency planning that is effective is able to provide effective and timely disaster relief operations (James and Paton, 2015).

There are three simple questions that can be used to break down the contingency planning process. These are:

- What will happen?
- What will be do about it?

- What are we able to do beforehand to make sure that we are well prepared?

These questions act as a guide that provides planners with the help they need to think through the plan in a systematic manner. We mostly use contingency plans when there are specific hazards or threats; the manner in which we will be affected by the threat remains unknown. A good way of thinking through possible impacts is by developing scenarios. Based on sensible scenarios, developing a plan that can set out the response scale as well as the needed resources is possible.

So as to be useful and relevant, contingency plans need to be an effort that is collaborative. It is also important to make sure that these are linked to the processes, systems or plans of other movement, partner or government bodies at all levels-global, regional, and national (World Bank Group, 2017).

5.4.8. Training in Disaster Management

Two of the four main areas for the focus on strategy 2010—the International Federation's strategy document—are disaster response and preparedness. The National Societies use this strategy as a guide in their work. It is also utilized by the delegations and secretariat that support them, by underlining the requirements for ensuring the improvement of the ability to respond to emergencies by the Red Cross Red Crescent Movement.

Disaster management training aims at building the capacity of the volunteers and staff in the National Societies, as well as that of the delegates in the International Federation, so as at to ensure the improvement of response and preparedness at all levels after, during, and even before the occurrence of disasters. This is all in addition to giving all of the Movement's components the means to work in a manner that is well coordinated.

The main focus of this training is to make sure that the participants attain improved technical skills; however, the training also improves the management skills of the team and the personnel. It aims at encouraging the exchange of knowledge and experience and also at creating networking opportunities for disaster managers. In addition, it aims at improving the disaster response coordination and the availability and quality of tools for disaster management (Figure 5.17).

Figure 5.17: Emergency management is important in avoiding the transformation of disruption into a disaster, which is considered much harder to deal with. Despite the fact that some people use these terms interchangeably, emergency management is related to disaster management; however, it is not equated to.

Source: http://www.tehrantimes.com/news/433489/National-document-on-disaster-management-training-prepared.

The International Federation has the role of making sure that the curricula is standardized, facilitate, and support trainings on disaster management at the international, regional, and national level, and also train trainers in National Societies. Red Crescent Societies and the individual National Red Cross have the responsibility of training at the national level, where they are supported as required by the International Federation technical delegates. The training normally reflects the National Society's mandate in the management of disasters in accordance to the country, and these vary country to country, but generally include:

- Water and sanitation;
- Volunteer management;
- Volunteer induction;
- Participatory hygiene and sanitation transformation;
- Community based first aid;
- Vulnerability and capacity assessment needs assessment training;
- Community based disaster management;
- IT/Telecommunications;
- Public health in emergencies;
- Relief and distribution;
- Basic training course for delegates;

- National disaster response team;
- Logistics management;
- Food security and nutrition.

At the regional level, disaster management training includes:

- Relief;
- Water and sanitation;
- Vulnerability and capacity assessment;
- Regional intervention team or regional disaster response team;
- Logistics management;
- Public health in emergencies;
- Emergency needs assessment.

At the international level, the training includes:

- IT/Telecommunications;
- Emergency response unit;
- Technical training in logistics;
- Team leader;
- Field assessment and coordination team;
- Emergency needs assessment.

5.4.9. Logistics Preparedness

When a disaster occurs, especially in areas that are remote and poorly funded, the response capacities that exist at the time may end up being invariably stretched. Apart from the operational issues that come up due to the geographical characteristics of a country, the capabilities of the national disaster response are also limited by:

- Especially in the remote areas, there may be poor logistics infrastructure. This may make it harder to provide effective responses. For example, lack of a road would prevent the response team from reaching this remote area.
- Vulnerable areas also lack facilities that they can use for the storage, handling, and consolidation of humanitarian cargo that they can later on distribute in areas that are affected.

This makes it evident that logistics preparedness is a very important component of any effort put into disaster reduction. Creating the plan is both practical and necessary, mainly because in general, we are able to foresee the disaster types that are more likely to affect a certain location, as well as the requirements that these locations will require to deal with the disaster. Resource assessment and vulnerability are the basis for logistics preparedness (Figure 5.18).

Figure 5.18: Adequately considering the issues that are brought about by logistics during contingency and preparedness planning greatly allow for us to achieve rapid mobilization of logistics operations in case of an emergency.

Source: https://www.ifrc.org/en/what-we-do/disaster-management/preparing-for-disaster/disaster-preparedness-tools/logistics-preparedness/.

Based on the results acquired after assessing the requirements of the affected areas, the Regional Logistics Unit that covers the affected area normally publishes a list of items that are for emergency relief, all which can be donated by National Societies as well as other parties. The Federation's logistics function manages the major role of coordinating the prioritization, planning the transport of, receiving, and distributing of the donated emergency supplies.

In the immediate aftermath of a disaster, the availability of a logistics stock preparedness can be very important in ensuring that the initial needs are met. There are emergency response stockpiles that have been planned, and these can feed a minimum of 300,000 people for a month. These have been located in a strategic manner in different cities globally, based in Panama, Lumpur, Kuala, and Dubai (United Nations, 2010).

For Logistics preparedness planning, there are some key questions that the planner can use to create a plan. It is a form of capacity and risk assessment. These are:

1. **The Disaster Type:** This involves determining the type of disaster that has a possibility of happening and making analyses on the impacts it may cause.
2. **Access or Entry Points:** This requires for the planner to be well informed on the various land, air, and sea entry points.
3. **The Transport Infrastructure that is Already Existent:** The planner is required to have adequate knowledge of the transport routes that already exist, and also use this information to find out how these routes could be affected by the occurrence of a disaster.
4. **The Available Resources:** This involves assessing and determining the capacity of the existing logistics.
5. **Defining the Logistics Needs Potential Response Scenarios:** This involves determining the logistics needs that may need to be added in case of a disaster that is in line with the scenario used during planning.

6

Environmental Disaster Recovery (Rehabilitation and Reconstruction)

CONTENTS

6.1. INTRODUCTION

This is in line with environmental management on issues disasters. Disasters can either be natural or caused by human activities. Most of the disasters experienced on the environment are due to the human activities. Most of the activities have led to disasters such as wild fires. The environmental management in matters disaster management aims at protecting the natural ecosystems as well as the natural environment. This management seeks to protect the environment. It sustains the natural environment, adapt to the effects experienced on the environment due to climate change, and reduce the disaster risks. The recovery is important to manage most of the disasters encountered in the environment such as pollution of water, desertification, and deforestation among others.

There is need for the recovery of the environment from such disasters to allow the sustainment of the world population. It is by far the most effective strategy used in attaining a sustainable environment. The recovery process is the role of the government as well as the citizens and should therefore be a combined effort to protect the environment. The environment needs to be protected from both the natural and the man-made disasters. Natural disasters refer to disasters that occur unexpectedly and there are no efforts that can be done to prevent them. They include storms, tornados, and earthquakes among others. There also man-made disasters are those that are caused as a result of human activities.

The environmental disasters directly impact the community. Take the case of drought occurring in various areas. It would mean that people would end up sleeping hungry and many would die due to hunger. Also, radioactive activities would lead to mutation and also harsh impacts on human health such as cancer among other diseases. Due to such destructive activities, the environmental resources have decreased causing the occurring migration of people to enable them adapts to the changes on the environment. The management of such disasters is therefore crucial for the sustainment of the future generation. Therefore, sustainment development which refers to meeting the needs of the current generation without any compromise being made on the current or future generation.

6.2. IMPORTANCE OF ENVIRONMENTAL MANAGEMENT ON DISASTERS

Environmental management on disasters is very important and given top priority. The reason is because disaster management facilitates the protection of the ecosystem by protecting the natural resources. It also facilitates a sustained development meaning that development of a given country can take place with no strain being placed on the environmental resources. It also reduces the risk on disasters occurring and allowing the environment to adapt to the changes in the climate. It also allows the government to look into methods of ensuring healthy environment for the citizens. This ensures that the citizens will have a high capability of being healthy. When environmental disaster management is ensured, it has the ultimate goal of ensuring the protection of the environment and health. This forms part of the major reason why environmental disaster management is important and should be carefully planned by the government (Srinivas, 2015).

In the event that a disaster occurs, the management which involve the humanitarian organizations coming to play. They are involved in the response which may be short term or long term phases of recovery. In most cases of disaster management, there are four phases involved. To some extent, the phases may overlap each other in the case of a cycle. The length of each phase is dependent on the severity of the disaster at hand. The phases are listed as shown below.

6.3. DISASTER HANDLING TIPS

6.3.1. Mitigation

This phase involves reducing the effects of the disaster. This phase involves building codes, zoning, vulnerability analyses, and public education. Preparedness. The phase includes the organization making plans on how to respond to the disaster. Examples consist of making preparedness plans, conducting emergency exercises, and warning systems. There is also the response phase. This involves making efforts to reduce the hazard created during a disaster which involves search, rescue, and emergency relief. That last phase involves the recovery phase which entails restoring the community to the normal situation. This phase involves construction of temporary housing, grants, and medical care (Srinivas, 2015).

In the conductance of the phases of disaster management leading to recovery, sustainable development is a variable greatly considered. They contribute all the aspects involved in the disaster management cycle. The sustainable development aims at facilitating sustainable livelihoods, protection of the environment and recovery in case of disasters and emergencies. In the case of a well-established sustainable development then there is a better chance of the state being able to deal with the disaster and therefore the recovery from the disaster will be faster and will be long lasting. In a development that is oriented around disaster management, it aims at minimizing the hazards, preventing disasters and general preparedness in the case of an emergency. The developmental considerations are greatly represented in the mitigation and preparedness phase of the disaster management cycle.

In the cases where the development processes are not appropriate then they have a high chance of being vulnerable to the disaster and the recovery process may take longer compared to a scenario where an appropriate development process is used (Srinivas, 2015).

This phase is composed of activities that involves the elimination and the reduction of the probability of a disaster occurring. It also involves activities that will minimize the effects of unavoidable/natural disasters. This involves measures such as building codes, zoning, and land use management, vulnerability analyses, building use regulations and safety codes, public education and preventive health care. This phase is highly dependent on the incorporation of the suitable measures used in both national and regional development planning. The level of effectiveness is dependent on the availability of the information concerning hazards, emergency risks as well as the countermeasures to be employed.

This involves the shaping of public policies and plans in a bid to modify the causes of the disaster or even mitigate the effects on people, property, and infrastructure.

6.3.2. Preparedness

The main aim of preparation programs used during emergencies is to attain an appropriate level of readiness to respond to emergency situations. This is attained through programs aimed at strengthening the technical as well as the marginal capacity of governments, communities, and organizations. The measures used during phase can be termed as logistical readiness in times of dealing with disasters. These measures can be improved by placing response mechanisms and procedures, rehearsals, development of both short-term and long term strategies, building early warning systems and public education. This phase can also be represented as strategic reserves of food, water, medicine, equipment, and other basic essentials are maintained in the case of a national or local disaster (Srinivas, 2015).

This phase involves individuals, organizations, and the government generates plans to ensure the safety of lives, enhance disaster response operation and minimizing the damage resulting from the disaster. It therefore includes making preparedness plans, instilling emergency communication systems, conducting emergency exercises and training, creating evacuation plans and training, emergency personnel and public education. The preparedness actions are dependent on the use of appropriate measures in both national and regional development plans. The effectiveness is dependent on the availability of information on emergency risks, hazards, and countermeasures to be employed. It also indicates the level to which government agencies; general public and non-governmental organizations are able to effectively utilize this information (Srinivas, 2015).

6.3.3. Humanitarian Action

In light of a disaster, the humanitarian agencies are summoned to deal with quick response and recovery. To allow them to effectively respond to disasters, the agencies are required to have capable leaders, trained personnel, appropriate communications, adequate transport, and logistic support and create guidelines on how to handle emergencies. If there were no preparations made, then the humanitarian agencies may not be able to meet the needs of the people quickly.

6.3.4. Response

This is an action taken in a bid to provide quick assistance to maintain life, improve health, and also support the morale of the affected regions. The response assistance involves a variety of actions that may range from giving specified limited aid. This is done through helping refugees in terms of transport, giving them temporary shelter as well as food, establishing semi-permanent settlement in allocated locations. They can also be assisted through carrying out repairs on damaged infrastructure. The response phase aims at meeting the basic needs of the people. This is done as the process of attaining more permanent and sustainable solutions are established. This phase greatly involves the humanitarian organizations being involved in this cycle of disaster management.

6.3.5. Recovery

This is the last phase in the disaster management cycle. It involves the disaster being brought under control. It involves the affected individuals being able to undertake a variety of activities that sought to restore their lives as well as the infrastructure they use. In the case of recovery, there is no defined point in which the relief become recovery and in the long run turn into long-term sustainable development. Various opportunities are presented in the recovery period to allow the prevention as well as increase the preparedness in light of an emergency. This reduces the level of vulnerability. In an effective recovery, it must consist of a smooth transition from recovery into on-going development.

Activities conducted during the process of recovery may continue for certain duration until the functioning systems are back to the normal state or a better one compared to the initial one. This implies that the recovery measures could either be short term or long term. The activities involve returning the important life-support systems to minimum operating standards. They include temporary housing, health, and safety education, public information, reconstruction, counseling programs and economic impact studies (Srinivas, 2015).

The response is also a major part of disaster recovery. The faster the response is conducted, the faster the environment is able to recover from the disaster. For one to identify the appropriate recovery procedure, they need to know the disaster and the impact it has had on the society. Therefore, allowing appropriate measures to be put in place to ensure a faster recovery from the disaster.

For instance, in the case of natural disasters such as avalanche which involves a drastic flow of snow down a slope, the response includes shutting off utilities. The recover process involves determining the impact of the avalanche on facilities and equipment. Then the government and organizations are able to take the appropriate measures in recovering from the disaster. In the case of a blizzard which is a severe snowstorm in most cases characterized by strong winds and low temperatures. When such storms have occurred, the immediate response may be power off equipment. The recovery process may involve listening to advisories and conducting an assessment on the damage and coming up with appropriate methods of assisting the people to come back to the initial phase (Srinivas, 2015).

In the case of earth quakes, the immediate response may be to evacuate buildings. The recovery process may involve the people setting up temporary structures for people affected by the earth quake. According to the intensity of the earthquake, buildings may be damaged. The recovery process may also involve providing the families with basic needs as the initial permanent structures are being set up. In wild fires, immediate response involves suppressing of the fires in the early stages. It may also involve evacuation of people as well as wild animals. As in the case of Australia where these fires are experienced, the recovery plans involve conducting assessments of causes of the fires and containing them. It may also involve creating temporary isolation areas where the wild animals can be kept. It also involves setting up mobile health centers where people can access health care services (World Bank Group, 2017).

In event of floods, the response may include monitoring of flood advisories, establishing emergency power generating equipment and damage assessment. The recovery procedure involves giving individuals temporary resting places or even establishing temporary structures for these individuals. Freezing rain may involve responses such as monitoring weather advisories. It may involve arranging for snow and ice removal. Hurricanes which are composed of heavy rains and high winds may involve responses such as listening to hurricane advisories, evacuating the areas, checking gas, electric lines and water damage. It also involves assessing damages caused by the hurricanes. The recovery process will be dependent on the type of damage incurred as a result of the hurricane.

Environmental disaster recovery is composed of a set of policies, tools, as well as procedure outlined to allow the continuation of important technological infrastructure and systems in line with a natural or man-made disaster. The disaster aims at keeping important aspects of the system to be functional despite the significant disruptive events. In the recent cases, there is the integration of IT skills in assisting in environmental disaster recovery. The assistance provided, is usually in form of a service. To further explain the integration of environmental disaster management, there are two case studies that were conducted to show the concerns in matters recovery (World Bank Group, 2017).

6.4. CASES STUDIES IN ENVIRONMENTAL DISASTER RECOVERY

The studies were mostly based on reviews of documents in relation to matters environmental operations. The documents include damage, loss, need assessment, and recovery plans. The reviews on the documents aim at looking in to the valuable lessons learnt in line with the disasters. They also look into the good practices involving the integration of environmental and disaster risk minimization. The cases selected during the study are those of Myanmar and Bangladesh. The environmental management community selected the two regions due to their experiences in dealing with environmental issues in processes while recovering from the disaster. In the case of Myanmar, the region experienced a cyclone that hit Nargis. The cyclone struck recently and the recovery process is still underway. Myanmar has demonstrated good practices in its planning as well as the assessment which has straight out proven and shows signs of a successful recovery. This implies that there is a higher chance of it recovering quickly and at a faster pace, leaving the environment greener than it was before.

Also, there was a cyclone that hit Sidr in Bangladesh. They were able to recover from this as they have an environment and long-term disaster risk reduction (DRR) strategy in place. This was possible as they had previous experiences when they had to recover from server flooding. The recent cases of flood event in 1988, 1998 and the recent being the Sidr cyclone. In the recent one, there have been a lower number of casualties that were recorded that what was expected. This is attributed to the fact that there have been improved disaster prevention measures that were put in place. They include improved forecasting and warning systems, cyclone shelters, coastal afforestation projects, and embankments (World Bank Group, 2017).

Both cases have however demonstrated sensitivities as they transition from the early recovery to recovery phase and present challenges that lead to the full integration of both environment and disaster risk management (DRM) minimization while in an early stage. With the strategies taken up by these two regions, it allows other states to learn from them and put in place better strategies.

Let us look into Myanmar. This is the largest country in south-east Asia having a total land area of 676578 sq. km. the country has a coastal line of 2,000 km and covers nearly the entire east coast of Bengal. This area is known to experience heavy rainfall in the mid-monsoon periods. This means that from June to august, they tend to experience floods regularly.

They are also prone to disasters such as cyclones, earthquakes, drought, and landslides. In the recent days, it is expected that they will face changes in temperature that may arise due to the climate change being experienced globally. The recent disaster they experienced was the cyclone that struck on 2nd and 3rd May 2008. The cyclone leads to a land fall in various divisions such as Yangon and Ayeyarwady.

The cyclones had wind speeds reaching 200 km/h. It was accompanied by heavy rainfall. Regions that mostly suffered due to the cyclone were the Delta as there were severe damages felt there. There were extreme winds also that were demonstrated by a storm surge that was 12-foot long. This cyclone was by far the worst natural disaster to ever strike Myanmar. It caused devastating impacts on the environment in the various divisions. This means that many lives were affected as majority of the population are dependent on the natural resources provide by the environment.

The damages were caused on natural resources as there was destruction of 35000 ha of natural and replanted mangrove vegetation. It also causes over 63% of the paddy fields being submerged and also about 43% of fresh water ponds were also destroyed. The impact of the cyclone was worsened by the pre-existing damages on natural resources. The damages include degradation and deforestation of the mangroves. It was also worsened by the over-exploitation of natural resources including fisheries. As the damages caused by the cyclone was heavily felt, it created a significant challenge in coming up with recovery efforts to allow the region to attain a sustainable development. One of the recovery tactics involves coming up with appropriate responses (World Bank Group, 2017).

6.4.1. Responses to Disaster Recovery Cases Studies in Bangladesh and Myanmar

The response activities involved combined efforts by the government as well as other organizations. The activities involved having coordinated relief efforts. The first response involved conduction a proper assessment on the cyclone to establish the impact of the cyclone on the environment. After this is done, they are required to come up with requirements useful in immediate humanitarian assistance as well as long term recovery. During the process, it was recognized that the environment plays a crucial role in ensuring the sustaining of livelihoods. Therefore, the natural resources will be a major part of the recovery process. The activities held during this process should ensure that the environment is able to recover at a faster pace. This lead to the great need of an Environmental Assessment being conducted. This was all outlined in the PONJA report which recommended the assessment to be conducted (World Bank Group, 2017).

The assessment is used as a basis for preliminary assessment of environmental damage as well as losses carried out. The assessment is guided by a thorough dedicated analysis and can be divided into four related areas. They include:

- **Direct Environmental Damage:** This involves the expansion and validation of preliminary findings on mangrove damage and is supplemented with detailed reviews of ground and surface pollution, sedimentation, salinization, and waste generation.
- **Environmental Footprint of Recovery:** This consists of relief and recovery activities that are to be conducted in an environmental friendly manner. It also involves the identification of additional damage to the natural assets and taking possible mitigation measures (World Bank Group, 2017).
- **Institutional Assessment:** This contains policy constrains as well as a capacity gaps at community, national level in managing environmental rehabilitation within a short to medium term. All these have to be identified.
- **Vulnerability assessment:** On long-term disaster and risk minimization strategy planning are undertaken. All these recommendations were outlined in the PONJA document. In a bid to response to the recommendations there was a project proposed to assist in recovery process used in Programme in Nargis. The UNEP is mostly involved in the realization of the project. The major objectives under the environmental assessments of the cyclones and risk management are as follows:

To develop a capacity in the National Commission for Environmental Affairs as well as related government agencies in an aim of renewing the environmental consideration used in recovery efforts in the affected areas.

The other objective is in the matte if long term assistance in the sense that the goal is to build a capacity within the NCEA as well as related government agencies to facilitate the integration of the environmental consideration used in both disaster risk minimization and management.

A three-year project was proposed with the objective of addressing the issues. The emphasis was placed on long-term recovery as well as the crucial role of environment and sound natural resource management in achieving sustainable livelihoods and promotion of disaster risk minimization. The project seeks out to cover capacity building as well as institutional strengthening in environment and natural resource handling while at the community, local authority and government levels (World Bank Group, 2017).

6.5. RECOVERY STRATEGY FOR ENVIRONMENTAL AND DISASTER RISK REDUCTION (DRR)

During the last decade, the impact of ecosystems in DRR has gotten increased universal attention. Sustainable ecosystem managing for DRR is presently recognized as a top priority aspect in the Sendai Charter for DRR, where UN Environment is part of the signatories and active members. Degradation of ecosystems-like forests, wetlands, marine, and coastal systems, plus drylands-is now broadly understood as a key driver of disaster risks and a major factor of disaster susceptibility.

6.5.1. Good Practice 1: The Placement of the Environment in the Strategic Framework

As was noted before, the environment plays a crucial role in the livelihoods of people. This is because it was noted that a majority of the population are dependent on the natural resources provided by the environment. This implies that in the case of a disaster occurring, the environment should be able to recover faster. This will ensure that the country will be able to come back to its initial level in matters development. The PONREPP is an alternative framework used by Myanmar. The advantage of using such is that it facilitates major considerations to be made in matters future preparedness. This is in addition to the existing emergency response. Therefore, the PONREPP provides a platform that allows a proper transition from an emergency relief and early recovery moving towards a medium-term recovery. This recovery is expected across eight operational sectors. It therefore covers a variety of themes the important ones being productive lives, protected lives and healthy lives (World Bank Group, 2017).

It was therefore very important to ensure that the environment was effectively included in the PONREPP. This applies for both the individual and separate sector. As an issue that cuts across a variety of subject matters, the environment was incorporated into the PONREPP under the theme on protected lives. This was a significant measure taken as the environment was initially seen to be relating singly to mangroves and therefore classified under forestry. There were several interventions made in the existing sectors and the results included the environment being included in to several sectors such as DRR, livelihoods, shelter, water, sanitation, and hygiene. Though this is significant progress, there are still some of the sectors that have not included the environment into their sectors.

The PONREPP team held discussions after which they decided to make the environment to be incorporated into the plan as a separate section. This is because among the responses from the discussion, most of the members felt that the chapters outlined in livelihood would not have been effectively sustainable without the simultaneous addressing of environmental concerns. This led to the suggestion that the office of the resident coordinator be used in providing a working group on environment. The project had to be implemented and the implementation process was started (World Bank Group, 2017).

The recovery strategy employed for the environment placed most of the emphasis on ensuring that sound management in matters environment and natural resource through use of strengthened systems at all the three levels (community, township, and national levels). Therefore, initiatives that were taken include a variety of activities such as the capacity building activities, institutional strengthening, systematic assessment of natural resources, strengthening of both monitoring and surveillance systems for the environment and natural resources and support for livelihoods-related schemes that are based on sustainable management of natural resources.

6.5.2. Good Practice 2: Explicit Environment for Disaster Risk Reduction (DRR) Capacity Building

Among the objectives of UNEP in terms of assisting Myanmar recover from the cyclone was to help them build a long-term capacity within both the NCEA and the relevant government agencies. This was to facilitate the integration of environmental considerations in disaster risk minimization and management. This involved a half-day session on environment and DRR being offered in programs such as training of trainers. The participants of this training were form townships within the affected areas of Nargis. The training was offered for the local authorities and it proved to be effective in demonstrating the existing connections between different sectors. There has also been substantial technical assistance.

This has allowed for discussions to be held with the WASH Cluster to facilitate training of trainers and equipment for water quality and safety. There are also some discussions being held with the department of land-use planning (Ministry of Agriculture) to facilitate the provision of equipment for the planned systematic surveys for soil salinity. There also other discussions being held such as the one with the Department of Forestry to take part in provision of technical assistance as well as equipment to allow them complete the assessment of mangrove resources in the Delta (World Bank Group, 2017).

6.5.3. Good Practice 3: Communicating Environmental Advice

Among the immediate response, after a disaster occurs (mostly natural) the major one involves seeking advice from specialists in that region. This therefore means that advice on the environmental issues should be at the disposal of the government and the relevant authorities. The advice should however, be requested and should provide a maximum impact. This means that question posed by the UN agencies on matters environment should be answered as required. Some of the case scenarios include the advice on IOM on the sourcing of materials to be used in rebuilding 300 dwellings. There is also UNICEF on environmental education for the curriculum they are developing for the Nargis-affected areas. Though this is made available, the government is advised to seek advice on matters environment if need be.

6.5.3.1. Long-Term Disaster Risk Reduction (DRR)

Most of the efforts taken are aimed at achieving a long term effect. There was a project concept generated on environment vulnerability and risk minimization in cyclone affected areas. This as well as capacity building to formulate plans and policies used in DRM. This involved the preparation of an initial draft that was done after a consultation with government officials, NGOs, and UN agencies. When the funded, the project would contribute to long-term preparedness of the state. The major goal of the proposed project is to facilitate the promotion of sustainable livelihoods and disaster management risk reduction in the Nargis-affected areas. This is achieved through the strengthening of systems used in sound management of environment and natural resources in all the levels (community, local authority, and national) (World Bank Group, 2017).

The goal can also be attained through the implementation of three-interrelated objectives that place emphasis on capacity building as well as institutional strengthening in all the levels having the community level being termed as objective one while the local authorities as well as the national government are termed as objective two. They are both supported by strengthening of systems for assessment and monitoring of natural resources in the affected areas termed as objective three. A better approach would be to assist in strengthening of ownership of activities done by communities and government agencies therefore helping to promote sustainability of the project in all the levels.

6.5.4. Good Practice 4: Scales of Entry Points for Environmental Management

As noted above, the recovery and preparedness plans were to be planned and initiated at the three levels. They include the community level also known as village level, the local authority level, and the national level. The three levels are later on termed to be the entry points. In Myanmar, the response is government based so these authorities (local) tend to be government agencies. These places the importance of the government being able to provide a framework for recovery and preparedness interventions. The interventions used in matters environment and natural resource management must place emphasis on the entry points. In the case of the environment and DRR as well as the PONREPP, the project concept aims at working at all the three levels. In the case of PONREPP, the focus is placed in the village level. This employs total participation in decision making, activity design, implementation, accountability, and management. Therefore, the interventions under PONREPP will place emphasis on villages and households. However, this is done under the consideration of the different needs of men and women (World Bank Group, 2017).

PONREPP also recognizes the role of the local authorities. This is because their level of governance is closer to the people. The presence of the local authorities in terms of active participation at township, district, and division levels in development initiatives. They are important for the coherent as well as equitable recovery and preparedness strategies across villages. They are also important infrastructure including schools, clinics, and community facilities are properly prioritized and approved in advance. They also ensure adequate environmental assessment and monitoring. This also applies for local capacity building and awareness rising on environmental as well as natural resource management is incorporated.

The representatives of local administrations and ministries at the township, district, and divisional levels are important in providing environmental and natural resource management by communities. They therefore play a crucial role in facilitating an enabling framework that support the community level initiatives more so in reference to access and control over natural resources. This will need a capacity building used for local administration in the implementation of the national rules and regulations. This is all in an effort of supporting the community-based initiatives (World Bank Group, 2017).

6.6. ON THE GROUND CHALLENGES

There are plans by the sectoral government which involves the incorporation of environmental elements. They are discussed in subsections.

6.6.1. Infrastructure

This involves the construction proper houses that have a higher possibility of withstanding disasters. This places emphasis on sustainable reconstruction.

- **Water:** There is need for clean water. The therefore look into increasing access to safe drinking water for the community.
- **Agriculture:** This involves the restoration of livelihoods as well as ensuring food security.
- **Fisheries:** There are plans to restoration of income as a result of fishing and subsistence fisheries.
- **Salt Industries:** Here, the salt fields are being rehabilitated. In matters forestry, there is the rehabilitation of forests and mangroves. They also include the preparedness and protection of the environment from future disasters (World Bank Group, 2017).

There are several lessons learnt from the disaster. They include the following:

- **Lesson 1: Communicating the Cross-Cutting Nature of Environment:** The past work placed emphasis on an analysis of government recovery plans. This is in terms of the incorporation of environmental concerns. However, these plans were hindered by the fact that most of the plans were written in Myanmar language and the translations were not readily available. Though this was the case, the contrary case was that on the basis of those available in English, there was an analytical framework that was prepared and two government counter parts are conducting an analysis on the available plans by making use of this framework. However, some of the government sectors still view the environment as exclusively part of the sectors they govern and have a preference of the environment being fully under their projects.

- **Lesson 2: Lack of Environment as an Initial Consideration:** The forestry plan was the only one taking the environment into its consideration in an outlined systematic manner. This is so though the implementation of the plan would need further environmental assessment is conducted before its execution. However, other agency plans have not taken to consideration the environment to any degree and more work is required for the analysis to be finished by the NCEA. This is when working with the agencies concerned.
- **Lesson 3: Funding and External Review:** Among the challenges commonly faced by the government is funding which is required for the implementation of the environmental plans. When allocation of funds is done, it would be beneficial to create an opportunity for the external technical review of the plans by the international agencies.

6.6.2. Conclusions of the Myanmar Case Study

Myanmar has made significant progress in recovering from the cyclone more so in the recognition of the contribution made by the environment in recovery and DRR. The good practices conducted in the case of Myanmar's recovery began at the choice of PONERPP instead of using a typical emergency response framework. This allows a better possibility in terms of transition from the response phase to the recovery phase. It also ensures that the environmental concerns from the beginning are integrated. By communicating the environmental issues to the suitable actors, providing the best training as well as technical assistance for both the cluster and the local authorities. All these improve the sustainability of the future environmental management with an aim of minimizing hazard risks. The capacity building explicit to the environment and DRR has vast momentum on all levels. However, the challenges are still a matter of concerning cross-sectoral cooperation as well as territorial ownership of the environmental issues. There is also a need to foster further governmental support for the issues in matters environment.

The second case is the one concerning Bangladesh. Bangladesh is a place known to be vulnerable to a variety of natural hazards such as cyclones, droughts, floods, and earthquakes. The result of these events is high numbers of deaths, considerable losses in terms of infrastructure and livelihoods. They also include a major setback in terms of the countries development. It has been quite difficult for the country to move forward as recently occurring disaster causes a major setback in terms of infrastructure and livelihoods time and again. There are several factors that have contributed to the vulnerability of Bangladesh in matters disaster. They include poor environmental as well as resource management, lack of environmental and infrastructure governance, rapid, and unplanned urbanization and population growth.

Though this has been noted, with the recent occurrences of the natural disasters, Bangladesh has been able to work on their responses and some improvements have been noted. This is since the 1998 floods which lasted for 10 weeks. A large percentage of the population was affected by it. In the case of the floods experienced in 1998, Bangladesh was relatively prepared compared to the one in 1988. This was though the areas affected were larger compared to the initial one and the effect of the floods lasted much longer. The improvements made on the response to the crisis were attributed to the increased credit access, economic growth, an increase in harvest as based on crop alterations, as well as a definite increase in preparedness through early warnings (World Bank Group, 2017). There are a variety of lessons learnt from the floods experienced. They include the following:

To begin with, there is the need to go beyond basic rebuilding: The environmental concern on matters recovery and long-term risk minimization was still a major area that requires more substantial consideration following the 1998 floods. Matters such as housing construction were identified as needing major improvement in the course of recovering, more so in the greater community participation as well as higher environmentally sound standards for design. In the case of cyclone Sidr that hit Bangladesh's south-west coast, the storm generated a high tidal wave of up to 5 meters and the coastal as well as the river surges were reaching up to 6 meters thereby breaching embankments and flooding of low lying areas. This cyclone has seriously affected multiple families. The strength on the winds was measured to up to 240 kilometers per hour (World Bank Group, 2017).

The floods experienced lead to serious damage of the infrastructure such as roads, bridges, housing, and sanitation. Some of the water resources became salinated or contaminated. The casualties were so many while some of the citizens reported to be lost while others are injured. Though the numbers recorded were relatively lower for the intensity of the cyclone. This is attributed to the previous effects made to improve the disaster prevention measures through forecasting and warning, coastal afforestation projects, embankments, and cyclone shelters.

The following are some of the Post-Sidr recovery strategies:

- **Good Practice 1: Integrating Environmental Management Principles at the Earliest Stage Possible:** Beginning with the early recovery strategy, the environment was given a distinct focus in the joint damage, loss, and needs assessment for the reconstruction and recovery after the Cyclone. In the joint damages losses and needs assessment (JDLNA), it is suggested that the environmental interventions should include an awareness raising campaign with the aim of preventing further damage to the vulnerable environment. There is also the activation of a sufficient geographic information systems/remote sensing (GIS/RS) based monitoring and evaluation system and reactive environmental rehabilitation and cleanup, restoration of damage environment and support to facilitate further protection, support for restoration of social forestry by employing the use of local plant species.
- **Good Practice 2: Explicit Building Back Greener and Better:** Initially, there was only partial environmental integration into recovery as most of the priorities were placed on credit and infrastructure assistance. The post-Sidr recovery strategy places its priority on building back better; enhancing the local capacity and utilizing environmental management. To provide local economic opportunities, activities such as rebuilding o houses, shelters, schools, rural roads, embankments, and markets. This introduces and mainstream new standards as well as upgrading that will assist in the protection against future disasters. It includes rehabilitation of the Sundarbans coastal forests ecosystems which are included as one of the Medium-to Long-term Recovery and Reconstruction Interventions. Some of the interventions include: construction of housing for a number of households, construction, and upgrading of transport network, rehabilitation of electricity services, rehabilitation of damaged or destroyed market places, reconstruction of urban public infrastructure,

reconstruction of water supply services, upgrading of health services infrastructure, and reconstruction as well as upgrading of damaged embankments.

- **Good Practice 3: Environmental Projects with Local Participation:** This involves the rehabilitation of the Sundarbans. On the basis of the evidence that the coastal forests of the Sundarbans. This provided efficient protection in the Cyclone Sidr as well as past events, the recovery strategy is as prescribed below:
    - Forestation along the embankments and have been integrated in the recovery program.
    - A long-term program that looks into handling matters concerning forestation along the coastal belt to minimize the vulnerability to storms and surges.
    - There is also the establishment of Large-scale rehabilitation of the Sundarbans. This project aims at the restoration of the Goral River system.

Beside the agenda, there was reorganization that the local capacity must be added to facilitate the maintenance of the afforestation and reforestation projects. By so doing, the environmental projects could assist in building resilience against future storm surge. They also provide long-term income opportunities residents in the form of future management of the forested areas. In the long run, it contributes to a better management of the resources therefore mitigation of future risks. The local participation in projects that is not explicitly linked to the environment. The advantage is that it allows the ownership and local inputs into rebuilding the needs. This helps the environment through a better long-term resource management thereby providing an understanding on new types of building materials are more resilient. This will later on avoid cases of wastage of the building materials as the population as they would have a better understanding of the type of materials they should use during construction (Srinivas, 2015).

6.7. RISK MITIGATION THROUGH EXPOSURE REDUCTION AND NATIONAL PLANNING

Risk mitigation refers to a strategy used to prepare for as well as reduce threat effects faced by data centers. Instead of planning to avoid the risk, mitigation focuses on the outcome of a disaster, plus the steps which can be taken before the event happening to diminish adverse, and possibly long-term, effects. The four kinds of risk mitigating methods are risk avoidance, acceptance, transferring, and limitation. Generally, the risks that must be avoided are those comprising of high prospect impacts for both economic loss and destruction.

6.7.1. Disaster as Part of the Development Continuum

In Bangladesh, flooding is a natural hazard that cannot be hindered. However, the vulnerability to the disasters can be minimized. Therefore, the focus shifted from the emergency response to long-term to sustainable livelihood protection is assisting to alleviate the losses in each hazard occurrence. So far, Bangladesh has been ranked to be the most vulnerable to climate in the world. For them to progress with sustainable development the DRM and reduction cannot be left out in the long-term planning. The opportunities need to be created in the case of a recovery to allow to be building back greener and better. There has also been a draft laid out in terms of disaster management. This draft is a good example in which Bangladesh is able to incorporate the environment and create the priorities in terms of dealing with climate change. The draft however, is still waiting for approval from the government (United Nations, 2010).

6.7.2. Environment Cannot Be the Only Answer

The risk management strategy employed in Bangladesh should be realistic as they are extremely vulnerable to climate hazards. The environmental consideration should be employed at all levels. This also requires them to identify that environmental protection may not be the only solution. In the JDLNA, it is pointed out that absolute protection in all areas against climate hazards is neither feasible nor economically defensible. The building resilience involves the combination of both soft and hard engineering solutions, zoning, and planning, innovative use of economic instruments in facilitating the protection of assets as well as direct incentives towards investments in climate-resilient forms of economic activity.

6.7.3. Environment Not Recognized in All Sectors

In the JDLNA, it was recommended that the funding for environmental need should be from the forestry department. This is attributed to the role it plays in the preservation of the Sundarbans. It is a suitable area for environment and disaster risk minimization, the recommendation risks losing a coherent and comprehensive environment recovery strategy. The proper allocation of these funds is really important. This is because the other sectors do not receive allocation for environmental matters. This would imply that they are less likely to make significant steps in matters integration of the environment.

6.8. EXAMPLES OF RECOVERY TOOLS FOR ENVIRONMENT DISASTER RISK REDUCTION (DRR)

Examples of recovery tools for environment disaster risk reduction (DRR) include the following subsections.

6.8.1. Risk Reduction and Environment (Integrated Risk Assessment)

It was developed under the international program for the International Chemical Safety. They both are jointly chaired by WHO, UNEP DEWA and ILO. The aim of an IRA involves addressing real situations of multi-chemical, multi-route, multimedia, and multi-species exposure.

6.8.2. Risk Assessment for Industrial Accidents (Including From Natural Disasters)

The awareness and preparedness for emergencies at local level (APELL) is a local process focusing on hazard communication and coordination. It entails a modular, flexible methodology tool using in assisting in the preparation in case of accidents. In an event that this fails. It aims at reducing their impacts. In the mission conducted within the framework of APELL has objective of addressing the risk exposure of local communities to certain events such as industrial accidents, tropical cyclones, flooding, wildfires, earthquakes, landslides, volcanic eruption, and tsunamis.

The APELL model has been used positively in many situations across the globe. APELL is a procedure "owned" and applied by local communities including their representatives. UNEP's objective is enlisting other global partners to ensure APELL is widely known, plus to help in its application through providing necessary information and advice.

- **Elements of APELL:** India has already approved APELL principles in countrywide legislation, including a series of both national and local seminars. A series of "safety-cells" have further been established in different communities. Besides, two huge industrial estates operating from China are currently adapting and implementing APELL as well. In Africa, Egypt has initiated a re-activation of the APELL systems at one of the large industrial complexes around Cairo. After some recent meetings, currently work is ensuing to establish a pilot-case-research of APELL applications in the highly industrial city (Figure 6.1).

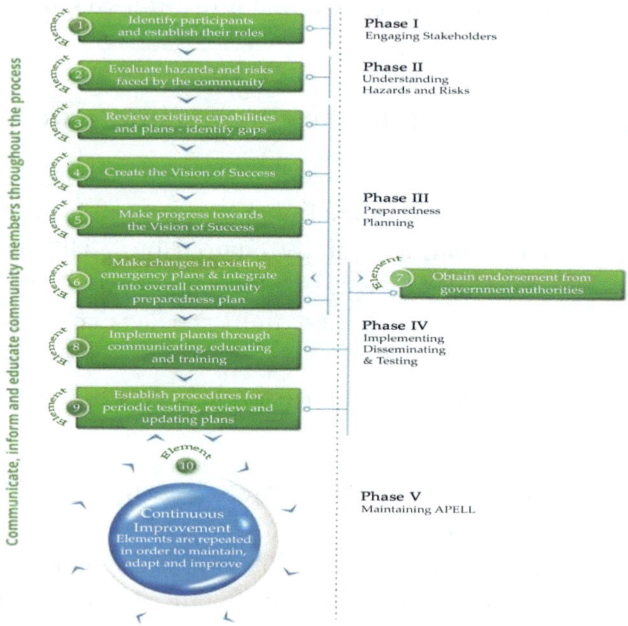

Figure 6.1: Examples of the application of APELL.

- **APELL Handbook (2015), Developed by UN Environment:** UN Environment established the 2nd version of the APELL Guidebook launched in 2015. The new edition points to the significance of an incorporated multi-hazard system at the local level, while emphasizing the significance of multi-stakeholder plus all-of-society involvement. The APELL framework is featured in a structured five phase and 10-element process, aimed at building a cohesive and robust community to both technological and natural hazards through increasing awareness and understanding on roles and obligations of community stakeholders towards developing systems for preparedness and disaster response.

6.8.3. UNDAC Disaster Response Preparedness Missions

UNDAC is designed to assist during the first phase of a sudden-onset emergency. The disaster response preparedness missions are conducted by the United Nations Disaster Assessment and Coordination teams in evaluating the countries' management as well as the response capacity and make recommendations for capacity building in this area. The evaluating capacity is important as it can reveal hidden assets and resources in the government agencies as well as civil society that can be built upon in an aim of strengthening preparedness and leading to a more predictable and efficient response as well as recovery process.

6.9. POST-CRISIS ASSESSMENTS

The aim of the Hazard Identification tool is to alert the UN country team after an occurrence of a disaster to potential secondary risk that are posed by large infrastructure as well as industrial facilities entailing hazardous materials found in the affected area. The information however can be shared by both the national and local authorities. This involves the addressing of secondary risk while in the early stages.

6.9.1. Strategic Environmental Assessment (SEA)

It is the role of SEA to ensure that the environmental consequences of plans and programmes undergo identification as well as assessment while in the preparation stage and also before, they are adopted. It is required that the public as well as the environmental authorities should give their opinion. The obtained results are then to be integrated and taken into accountability in the course of the planning procedure. When the plan is adopted of the plan or the programme the public gets to be informed about the decisions made and the manner in which they are made. If there is a significant trans-boundary effects, the affected Member State and the public is informed and are given on the chance to make comments. The comments can then be integrated into the national decision making process.

6.9.2. Environmental Impact Assessment (EIA)

The procedures outlined aim at ensuring that the environmental consequences of projects are established and assessed before an authorization is given. The public is given the opportunity to give their opportunity and the results are taken into consideration in the authorization procedure of the project. Afterwards, the public is given information of the decisions made.

6.9.3. State of Environment Reporting (SoE)

This program refers to the relevant conditions of the region in the view of two perspectives: that is the socio-economic as well as the bio-physical conditions and trends. This report aim at seeking to address: emerging issues in the region, existing policy responses at national, present environmental status and trends, sub-regional, and regional level; future perspectives on the basis of past and present trends in the different patterns and the recommended policy action. This reporting is aimed from the grass-root to the high-level decision makers. A good example of SoE in a post-crisis location is for South Sudan (2018). A state-owned enterprise or (SOE) is a legal body established by the government to participate in economic activities on the state's behalf. In South Sudan, the oil industry largely follows this model.

6.9.4. Flash Environmental Assessment Tool (FEAT)

The provision of a rapid scan to identification of the most acute environmental issues is catered for by the FEAT. It allows immediate response once a natural disaster occurs. It (FEAT) has its major priorities placed on issues that arise due the released chemicals. It gives off general indications of the type of impacts that are expected from physical occurrences. They include erosion of fertile soil and salt water intrusion. This is incorporated as part of the FEAT as a desktop screening.

6.9.5. Emergency Waste Management Guidelines

These guidelines were developed by OCHA as well as the Joint Environment Unit UNEP/OCHA. They give advice to the actors conducting emergency response on good waste management practices such as initial clearance, storage, and disposal.

6.9.6. Strategic Assessment (SA)

This involves the provision of a means of undertaking a better integrated response therefore allowing senior decision makes to establish the appropriate form of United Nations engagement. Their object does not involve the repetition of previous assessments or the validation of ongoing programmes. Their aim to indicate possibilities of the United Nations being able to reduce coherence, impact, and focus. One of the post-crisis assessments is an Integrates Strategic Environmental Assessment (SEA) developed by the UN.

During the early periods following conflicts or disasters, it's essential to guide the preliminary steps of recovery or reconstruction, so that fresh developments reduce environmental impacts and create resilience against disaster, climate, or conflict risks. Post-Crisis Integral SEA is a technique that arose from the post-crisis plus development process of Northern Provinces of Sri Lanka, following 33 years of recurrent conflict.

6.9.7. Post-Disaster Needs Assessments (PDNAs) and Post-Conflict Needs Assessments (PCNAs)

These groups are jointly part of the UN-EC-World Bank missions conducted to facilitate the production of basic post-crisis assessment report. This makes use of the sectoral PDNA methodologies that were developed by specialized agencies. The objective is to identify the priority areas as well as the financial requirements needed for post-crisis recovery as well as the reconstruction.

6.9.8. Post-Conflict Environmental Assessment (PCEA)

This is a tool utilized by UNEP to facilitate an objective scientific assessment of the environmental situation experienced in a country after the occurrence of a conflict. The objective is to educate the general public on environmental risks related to the conflict. They also seek to provide guidance to governments on priority to the addressed issues.

6.9.9. Joint Damages Losses and Needs Assessments (JDLNA)

This was generally led by the World Bank Global facility used in Disaster Risk Minimization. This has an objective to identify recovery needs and quantify them.

6.10. POST-CRISIS RECOVERY

Post-crisis recovery is among the least addressed step in planning, training, or simulations. It's an aspect that, when not properly managed, may cost economically, reputationally, and operationally. Guiding principles for post-crisis recovery usually are lacking; besides many entities seem lose focus in terms of discussing post-crisis recovery plans. It could be that post-crisis recoveries are among the most complex of Business Continuity Lifecycle factors, plus that no two recoveries can follow a similar pattern. Nevertheless, post-crisis recovery processes can be subdivided into manageable parts that may be done through a project management technique.

6.11. COMMON COUNTRY ASSESSMENT (CCA)

According to the definition by the General Assembly, this is a basic instrument of the UN system to conduct the analysis of the national development situation as well as identifying important development issues. The CCA takes into consideration national priorities focusing on the MDGs as well as other commitments, targets, and goals of the Millennium Declaration and international conferences, conventions, and summits.

There are multiple measures being taken into building back a greener and better environment after a disaster occurrence. This is through the implementation of sound environmental practices that will lead to the minimization of future disaster risks. However, some of these are not available. It creates the need for the conductance of a further detailed analysis seeking to identify ways of optimizing environmental issues in donor policies and national risks. Further case studies have been conducted with a systematic comparison of the best lessons learned as well as the practices are required to assist in the answering of questions such as what constitute environmentally sound relief and recovery operations, how the environmental actors are engaged in disaster recovery and also what support and guidance in relation to the environment being availed in disaster context (United Nations, 2010).

Therefore, the future works require further detailed and better pathways that will facilitate the recovery community and the best measures need to integrate environment as well as disaster risk minimization post-crisis. This will involve early stages of recovery with the aim of creating a seamless link between recovery and response phase inclusive of the short, medium, and long-term environmental issues that are collectively addressed in an integrated manner. It is crucial in distinguishing the needs in the different recovery phases. In particular situations however, there are no capacity on the ground in the early recovery till the implementation of the recommendations.

In the case of an extensive study of lessons learned and good practices, there are proposed indicators used in the reviewing of the integration of environmental concerns used in recovery operations:

6.12. INCORPORATION OF ENVIRONMENTAL ISSUES IN POST-DISASTER NEEDS ASSESSMENTS (PDNAS)

This involves a variety of activities such as the identification of environmental factors known to contribute to the disaster, priorities for reduction of environmental vulnerability. This also includes other environmental safeguards used in recovery and reconstruction. The assessments involve the use of impact data in demonstrating the effectiveness of environmental risk minimization such as comparison of damage as well as the losses in the selected areas having contrasting environmental quality or resources bases. In this case, the highlighting of gaps is equally important such as disaster waste management. This is a common issue in large scale natural disasters which are not always addressed. It is necessary to document the lessons learnt in relation to matters transition from rapid environmental assessment to early recovery planning as well as programming. They include the review or related guidelines in the given topic.

6.13. THE EXTENT TO WHICH RECOVERY PROGRAMMING AND PROJECTS HAVE CONSIDERED OR NEGLECTED ENVIRONMENTAL CONCERNS

The presence of environmental projects in post-disaster recovery plans as well as the mainstreaming of environment in sectors such as shelter and sanitation among others should be put under consideration. Some of the case scenario is such as post-disaster responses that have contributed to the inadvertently exacerbated present environmental weakness.

6.14. EVALUATION OF THE IMPLEMENTATION OF RECOVERY PROJECTS

It is important that the success and extent of implementing of measures taken in environmental DRR. This achieved through the documentation of completed projects as well as resources that have been employed. This involves the cases involving disaster that have resulted to a retreat in areas having inappropriate development environmentally. It also includes the determinants of an accomplished retreat including quick decision-making structures and preparedness. To facilitate the provision of opportunities used in building back greener and better, post-crisis events should be taken into consideration. This implies that they should be taken into consideration in light of a future study. The interrelationship present between the environment, recovery, future conflict prevention, as well as development is important and therefore requires more attention.

A matter given great emphasis is funding. It is important that the environment is given more priorities and also ownership is taken up in all the sectors. This will facilitate the maintenance of funding. If the funding falls short, then the result will be that the environment as well as the DRR opportunities in the build back greener and better will be foregone as there will be insufficient resources to support it. There will also be inadequate information to the donors greatly involved in making decisions with regards to the funding. This will imply that in long-terms some countries such as Bangladesh will not be able to afford certain opportunities that will lead to the minimization of risks. It is therefore important to note that the best investment involves acting on the knowledge on the lessons learnt as well as good practices obtained from past recovery planning. Also the operations used in facilitating optimized recovery in each disaster occurrence in line with the long term development continuum.

6.15. ORGANIZATIONS INVOLVED IN ENVIRONMENTAL DISASTER RECOVERY

There are several organizations involve in the recovery processes in countries that have encountered a disaster. For instance, the United Nations Office for the Coordination of Humanitarian Affairs is the overall body that is in charge of other organizations playing a significant role in this process. Its major role is to bring together all the humanitarian actors. This is done with the aim of attaining a faster response when such disasters strike a country. It therefore plays an important role of coordination operations during such crisis. They conduct a series of activities such as assessing the situation as well as the needs, coming up with strategies useful in addressing the challenges at hand, mobilizing of the funds, and also giving the public information. Depending on the country, there is an organization that is in charge of handling situations such as natural disasters (Figure 6.2).

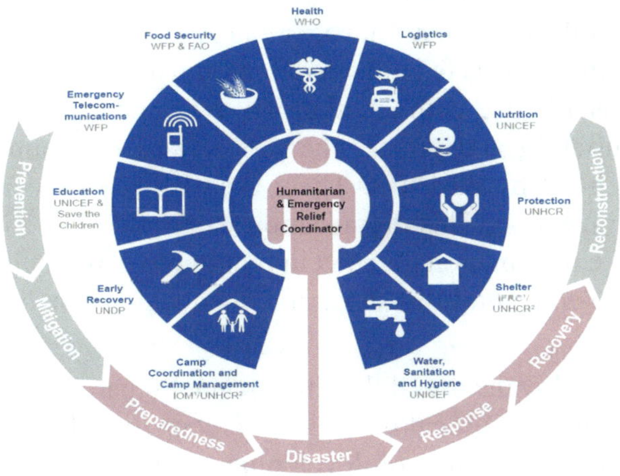

Figure 6.2: The UN lead at various phases of a disaster.

Source: https://www.unocha.org/sites/unocha/files/1823826E_interactive_menu.pdf.

For instance, in the United Kingdom, the organization used is RAYNET. The organization provides response through the use of off-road vehicles. In Canada, the organization in charge is Global Medic. It was established as a non-sectarian humanitarian aid that seeks to provide disaster relief services in the case of larger scale disasters. This organization is composed of volunteers who take part in the response. They take on activities such as professional rescuers, firefighters, paramedics, and police officers. This organization is more organized as they have equal distribution of tasks. For instance, they have a Rapid Response Team that offers its services in the rescue units. There are also the water purification units (WPUs) that provide clean water that is safe for drinking as well as the emergency medical units (EMUs) which make use of the field hospital in the provision of emergency medical treatment (World Bank Group, 2017).

In India, the organization in charge of the response is the National Disaster Management Authority. The body is in charge of conducting mitigation effects of natural disasters. They also assist in avoiding man-made disasters. They also conduct capacity building as well as response of the government in cases of these disasters. They also make use of the National Disaster Response Force which is an international agency in charge of handling the disasters. In the US, the response and recovery activities are led by the Federal Emergency Management Agency (FEMA). This agency conducts activities such as federal operations and logistical disaster response capacity. This is all done with the aim of saving and sustaining lives. They also reduce suffering and protect the property of communities that tend to be overwhelmed by disasters. There are also the Centers for Disease Control and Prevention. This organization works by giving information to the citizens. The information given is in relation to certain disasters such as disease out breaks, natural disasters, and chemical and radiation accidents among others.

They also employ the use of the Emergency Preparedness and Response program that is of the National Institute for Occupational Safety and Health. This organization is in charge of developing resources useful in addressing responder safety and health employed during responder and recovery operations. The US also employs the services of the American Red Cross in conducting non-profit efforts. This organization is supported by religious groups as well as community service agencies. Though the government plays the major role in responding to the disasters and aiding the recovery process, there are also non-governmental agencies that take part in these activities. There also some charities that assist the government through the provision of funds. The main organizations involved in the disaster recovery are under the International Umbrella Organization. Some of the organizations under it include the Inter-Agency Standing Committee and the International Council of Voluntary Agencies (World Bank Group, 2017).

From the perspective of the two case studies, it has been noted that for a country to recover from a certain disaster, there is need for the identification of the disaster. Once the disaster has been identified, then the appropriate response and recovery measures can be undertaken. Though most of the disasters tend to have the same responses practices to be undertaken, there are some measures that are specific to a given disaster. For instance, in responding to a disaster in drought and on the other hand handling disaster on flooding, the measures undertaken may not be the same. In this case, the response measure will be specific to the disaster at hand. In response and recovery of a given disaster, there are technologies that are used to aid in the process. These technologies are usually identified under the emergency management software.

They include the use of a Smart Emergency Response System which was an initiative of the United States. This initiative seeks to challenge individuals to build cyber-physical systems that can be used in the future to save lives, foster businesses and build the economy among other roles. However, its major goal is to save lives. This system works by giving information to the emergency personnel as well as the survivors so that there is easy location and assistance amidst the disaster. It creates a platform where the survivors and victims are able to relay their help requests to the emergency response team who in turn will give the required assistance. There is also an International Charter Space and Major Disaster. This board assists in disaster recovery through the giving of satellite coverage. This allows the involved personnel being able to easily identify the affected areas. It may also be useful in giving an in-depth analysis of the level of damage incurred by the environment (Babar, 2007).

Environmental disasters are man-made. Most of these disasters have led to the extinction of certain species of plants or animals. In cases where the human activities are the ones contributing to the disaster, to facilitate the recovery of the environment, the activities are conducted under supervisions or the activities are halted. For instance, mining has led to destruction of most of the parts of the environment. This is because after the mining activities are conducted and completed, the land is left bare and is left unproductive. To allow the recovery of the environment with relation to mining, the measures under taken include restoring of the land in its initial state. The government also gives a directive on how the activities should be conducted and the measures that should be undertaken. For instance, in mining, it is advocated that trees should be planted in the areas that mining was once conducted. It is a true realization of the fact that according to the type disaster caused, there is a measure to be used that will aid in recovering of the environment (Birkmann, 2007).

7

Environmental Disaster Mitigation

CONTENTS

7.1. INTRODUCTION

Mitigation refers to the steps taken to inhibit or lessen the threat to life, property, socioeconomic activities, as well as natural resources coming from natural or manmade hazards. Some key measures that can be considered include; creating more awareness, preparedness, education, and early prediction and warning structures that can lessen the disruptive effects of natural disasters on communities (Babar, 2007) (Figure 7.1).

Figure 7.1: An oil spill disaster polluting ocean waters.

Source: https://safetymanagement.eku.edu/blog/disaster-mitigation-4-oil-spill-responses/.

Moreover, mitigation measures like applying zoning, land-use policies, and development of codes are required, for preventing or reducing the extent of destruction from hazards. For instance, avoiding construction in landslide- or flood-prone regions through proper preparation and zoning regulations, may save finances in the building process and minimize loss of life, damage to property as well as natural resources. Besides, post-disaster studies continue confirming the basic fact that community participation in mitigation has direct dividends in curbing disasters (Babar, 2007).

7.2. EMERGING TRENDS

In the U.S., development practices nowadays emphasize the need to impart a fresh commitment to disaster mitigation. The National Institute of Standards and Technology (NIST) estimates a countrywide investment of trillions of dollars in fresh construction and infrastructure.

Advancements in the sectors of hazard or risk evaluations are also giving decision-makers increasingly factual and useful data on the possible effects of natural calamities on society. But the main challenge is applying this data and translating existing understanding of natural threats into meaningful activities at community level, protecting substantial investments in fresh and existent developments. Just as significant is the human aspect of mitigation; programs must highlight the needs for urban, aging, and technologically focused society (United Nations, 2010).

In spite of increased susceptibility to natural calamities, many communities still resist embracing mitigation programs. The obstacles are socioeconomic, political, plus mitigation is usually considered as restrictive, expensive, and not compatible with the society's economic development objectives.

Mitigation may involve programs that are technically rigorous but politically not accepted. In this perspective, elected officers are usually reluctant to follow mitigation measures vigorously. Any progress toward adopting the mitigation practices shall require community participation, realization of constraints and obstacles, and inventive solutions. Flood-prone regions, for instance, may be integrated into community-improvement open space, recreational, and wildlife attractions, physical fitness, or hiking trails. Besides, new awareness and educational programs must be developed to promote integration of lessons learnt from disasters, plus various findings from social studies into mitigation practice (Elliott and Hsu, 2016).

Mitigation measures, above all, must involve the primary groups that contribute to development, adoption, implementation, and enforcement mitigation-public officers, finance, and insurance professionals, engineers, developers, and architects, civic societies, marketing experts, educators, and scientists among other stakeholders.

For maximum effectiveness, mitigation needs a multi-disciplinary team strategy that's free from control by any particular interest group; every discipline has some objective and input to make. Furthermore, close communication and synchronization among scholars, practitioners, and legislators increase the probability that effective mitigation measures will be implemented.

One popular program for improving hazard mitigation capacities must capitalize on prospects for instant and near-term success, plus sustain development into the coming years. Many affordable mitigation programs must be determined and profiled, before the best possible solution is identified (Smith, 2003).

Besides, transferable solutions should be closely inspected. Similarly, a balanced Decade project must address the basic problems related to mitigation—, socioeconomic, and political obstacles. The U.S. Disaster Committee suggests that every community which is at substantial risk must adopt and implement a proper mitigation plan, covering both short-term goals and a detailed long-term plan for curbing the effects of natural disasters.

In order to realize this objective, the U.S. Disaster Committee recommends that local authorities take the below steps: integrate both operational and nonoperational mitigation procedures in fresh development, inspect ways to lessen the susceptibility of existent structures, take steps towards reducing the susceptibility of natural resources, plus undertake mitigation teaching activities with help from state and the federal administrations.

Moreover, a program for improving countrywide hazard mitigation abilities must include: Protecting schools and hospitals. Every new school and hospital must be located is a safe place, and built to ensure high-hazard zones are avoided plus that unique provisions are provided to reduce the possibility for destruction by natural hazards. Additionally, existent school and hospital structures must be surveyed to decide their levels of resistance to possible hazards (Smith, 2003).

Consistent efforts must further be made to reinforce facilities which are highly likely to fail when disaster strikes. For some cases, legislation might be needed to ensure that general mitigation measures are taken. Additionally, disaster mitigation experts believe that emphasis should be placed on applying mitigation programs in schools as well as hospitals due to their essential part in community life, plus their intensified significance during disasters. Among the country's greatest resources—who are kids—spend a large amount of time at school. Plus, schools also work as main shelters for evacuees in times of disaster. Oftentimes, though, school structures are neither developed nor maintained to fully withstand the physical outcomes of natural hazards, plus in many states, they're actually exempt from construction codes. Therefore, they are possible death traps for pupils or evacuees inside them. Hospitals including other healthcare facilities attend to the sick or injured, plus are the center of medical technology as well as expertise that's crucial in disaster times (Simelane and Thokozani, 2019).

Whenever hospital facilities fail following a disaster, like they have in many earthquake and hurricane situations, not just are patients or medical personnel slain or trapped inside them, but affected communities are furthermore deprived of basic medical resources, supplies, and equipment.

When economically feasible, communities must strengthen other potential facilities like emergency operations facilities, police or fire stations, amenities, and telecommunications as well as transportation networks, these are crucial to emergency responding and recovery. Additionally, convention facilities, theaters, museums, and other areas of public assembly must also be integrated in a process to strengthen buildings. Procedures are also required to integrate the private industry in mitigation measures for both the public and private buildings (Simelane and Thokozani, 2019).

7.3. ADOPTING NONSTRUCTURAL MEASURES

Companies and private homes must integrate nonstructural mitigation procedures to reduce injuries, and property destruction from natural disasters. Furnishings and equipment, for instance, may be easily secured in order to lessen injuries and destruction from earthquakes. The other non-structural procedures are managing of vegetation in order to minimize damage from wildfires, plus location of vital structures far from high-hazard zones (Birkmann, 2007) (Figure 7.2).

Figure 7.2: Remains of a burnt car following a bushfire outbreak.

Source: https://www.cfr.org/backgrounder/us-disaster-relief-home-and-abroad.

Nonstructural mitigation signifies a major chance for instant low-cost action to diminish the effects of natural risks on the residence and the workplace. Furthermore, the private sector may contribute considerably to promoting non-structural mitigation. Besides lending institutions are normally positioned to integrate mitigation procedures as stipulations for loans, plus the insurance and re-insurance industries may adjust underwriting rate programs as an inducement for mitigation (Birkmann, 2007).

7.4. INTEGRATION OF MITIGATION MEASURES INTO NEW DEVELOPMENT

Regional jurisdictions must ensure that any upcoming development is located, developed, and developed to resist natural risks. They must use information from danger and risk evaluations, land-use policies, and zoning procedures to minimize development in disaster-prone areas. Besides, compatible applications of floodplains and related hazardous zones must be integrated into regional planning and zoning, such that losses may be reduced. These areas might have a great value for recreation, open space, as well as marine and wildlife reserves (Weil, 2004).

Development codes that establish minimum conditions for development and construction may be a powerful instrument for moderating the impacts of natural risks; many lives can be saved when proper mitigation measures are put in place to avoid a disaster. Some of these include; dam safety standards, emergency provisions, and fire codes among others.

Furthermore, building codes must be a crucial part of any mitigation measure for new construction. Obstacles towards the adoption and implementation of modern codes must be determined and approaches established that can help communities cope with and overcome possible disasters (Weil, 2004).

Additionally, even though land-usage plans, zoning regulations, and building codes plus regulations mostly are considered the obligation of state and local governments, still financial aid and technical support will be required to adopt and apply these mitigation policies. The assistance can be provided through agencies like FEMA, NIST, and professional companies like the Applied Technologies Council, Building Seismic Safety Committee, and the Earthquake Engineers Research Institute.

7.5. PROTECTING CULTURAL PROPERTIES

The protection of libraries, historic buildings, monuments, artworks, and other culture based resources must be integrated into mitigation planning systems and action. In California, Hurricane Hugo which hit Charleston district was responsible for destroying cultural properties which were particularly vulnerable.

There's need for both data dissemination and practical assistance for curbing this type of loss. After a disaster, the preservation of historic venues may be an emotional, plus costly element of recovery and restoration. Mitigation training furthermore must include the subject of preservation for promoting informed decision-making as well as community involvement (Veenema, 2003).

Both national and global groups are currently working under the guidance of UN's Educational, Scientific, and Cultural Organization (UNESCO), International Council of the Museums, plus other organizations for raising awareness of the risk of natural risks to cultural properties plus to alleviate or prevent possible destruction and losses. Besides, foundations, and private bodies may contribute to these practices.

7.6. SAFEGUARDING OF NATURAL RESOURCES

Especially valuable natural resources like endangered groups of wildlife, plants, and fish must be detected in mitigation policies and protection measures applied like disaster response plans. These natural resources are present not just in the wild, however in parks and zoos also.

Mitigation plans may include particular attention towards the location and designing of facilities, so that the fire or windstorm doesn't function as a channel for unexpected destruction to essential natural resources. For instance, pipelines, and power grids regularly traverse crucial natural resources zones. In such circumstances, it's possible to expect probable destruction to nearby natural resources, which are caused by rupturing of the pipeline or wrecked power grids (Veenema, 2003).

Similarly, special breakers, programmed flow controls and different other features are easily available and may drastically diminish damage. For especially valuable and endangered groups of plants and wildlife, practical planning may include transferring a part of the population, so that natural events don't cause any loss to the population (Veenema, 2003).

7.7. STATE LEADERSHIP OF MITIGATION OPERATION

The leadership or administration at every level must set an example through requiring that newly built facilities which it finances, regulates, or leases should be designed, developed, and located according to modern construction codes and comprehensive land-use practices (United Nations, 2010).

For national level, in January 1990 the U.S. government provided an executive order needing those federal agencies should develop and construct different buildings in order to become earthquake resistant. Other similar directives, reinforced by active enforcement projects, are required at the both state and local ranks, besides they must encompass all applicable natural hazards. These standards must be taken into consideration for publicly-funded infrastructures, plus lifelines like highways and bridges. When resources permit, it's also necessary that requirements must be extended to all existing buildings within a phased program which highlights their vulnerability (Cutter, 2012).

7.8. MITIGATION TRAINING

The training programs that concentrate on modern challenges related to implementation of mitigation measures must be well-developed and offered. Furthermore, a national training project, supported by federal government or fully incorporated with preparedness of training must be established for this reason. Its curriculum may cover topics such as land-use planning, tax incentives, non-structural mitigation measures, zoning, construction codes, and regulations. Furthermore, case studies from all around the U.S. world must be incorporated. Mitigation training programs provided by USFS and FEMA must include more contributors, disciplines, and subjects (Cutter, 2012).

Furthermore, mitigation training must be highly collaborative, while revealing real problems and concerns. For instance, it must be determined how hazardous and risk data can be used to enhance mitigation at the communal level.

7.9. HAZARD-FOCUSED RESEARCH

Many recent disasters have shown the advantages of mitigation efforts, while indicating the need for science to enhance mitigation practice.

Even though various hazards can benefit from these studies, still research agendas touching on earthquakes, extreme winds, and landslides need to be strengthened.

For earthquakes in particular, there's need to successfully complete a nationwide seismic monitoring network, plus establish a supportive global program in strong-motion evaluation and data analysis. Additionally, local networks must be structured, as needed, to regulate the effects of regional site situations on ground motion, and the connection between particular ground motion stipulations and extent of structural damage (World Bank Group, 2017).

7.10. THE POLLUTER PAYS PRINCIPLES

'Polluters pays' principle is a widely accepted practice where those who pollute must bear the rats of managing it for preventing damage to people's health or environment. For example, a factory that generates a potentially poisonous element as the byproduct of its activities normally is held liable for its safe discarding. The principle reinforces most of the guidelines of pollution concerning land, air, and water.

7.11. MANDATORY ENVIRONMENTAL INSURANCE (ENVIRONMENTAL LIABILITY INSURANCE/FUNDING)

In many jurisdictions, insurers have become gradually concerned around pollution liability risks. The insurance markets theoretically can allow effective sharing of environmental damage rates and offer incentives to prevent environmentally reckless behavior. Some experts also believe that contribution of insurance towards these policy objectives can be improved when environmental obligation insurance is required by the state.

7.12. BUFFER ZONES

Buffer zones refer to a tract of land covering two differently zoned zones. For instance, a city may place a park between the commercial and residential quarters. In Sub-Sahara Africa, management of buffer areas around protected zones doesn't usually take into consideration the requirements of resource exploitation through the local populations, or conservation requirements of these areas.

7.13. ENVIRONMENTAL IMPACT ASSESSMENT (EIA)

Environmental impact assessment (EIA) is a procedure for evaluation of possible environmental impacts of the projected project or development, considering inter-related socioeconomic, cultural, and human health effects, both advantageous and not. The role of EIA process is notifying decision-makers and members of the public on environmental effects of applying a suggested project.

8

Environmental Health Management

CONTENTS

8.1. INTRODUCTION

All aspects of our natural environment and that which is made including the air we breathe the water we take, the places we work or interact and the food we eat may have a direct or indirect impact on our health. Interactions between human genetics and the environment or our physical interactions, which is termed as complex can bring about the occurrence of an array of health or disease conditions. Once this happens, it becomes a public health concern and this has led to the development of a branch of public health referred to as environmental health. Over the years, the incidences of diseases of certain diseases have increased and linked to the built environment rather than the natural environment as a result of changes man has made to the environment. Hazards within the environment are known to increase disease risk. A variety of environmental exposures can be related to effects to the human health for example:

- The environmental exposure to polluted air can lead to cardiovascular diseases, respiratory conditions and cancer of the lungs.
- Exposure to unsafe or contaminated drinking water can lead to gastrointestinal diseases such as typhoid, cholera, shigellosis, and dysentery among others.
- Recreational contaminated water exposes one to gastrointestinal diseases and infections of the throat, eyes, ears, and the nose.
- Exposure to disease vectors such as mosquitoes can lead to diseases such as malaria, dengue fever, and yellow fever.
- Frequent exposure to ultra violet radiation can lead to melanoma, cataracts, and non-melanoma skin cancer at the same time when there is little or reduced exposure vitamin D deficiency, rickets, osteoporosis, and osteomalacia can occur.
- Exposure to smoke can lead to ischemic heart disease, stroke, or lung cancer. In infants, passive exposure can cause low birth weight and sudden unexpected death in infancy. In children, smoke can lead to respiratory tract infections such as asthma.
- Crowding in households can lead lower respiratory tract infections and also lead to the spread of infectious diseases. Living in housing conditions that are cold and damp can lead excess morbidity rates.

- A change in climate can result in an increased occurrence of infectious diseases and heat stroke.
- Exposure to substances that are hazardous can lead to poisoning, burns, dermatitis, and consequently death.
- Exposure to heavy metals such as can lead to developmental delays in children and problems with behavior. Exposure to asbestos can lead to difficulty in breathing, cancer of the lungs and mesothelioma.

According to the World Health Organization report published in 2016 there are an estimated 12.6 million people who died as a result of working in unhealthy environments in the year 2012, which is nearly one of four reported deaths. Some of the environmental risk factors identified as the causes of diseases included air, water, and soil pollution, exposure to chemicals, changes in climate and ultraviolet (UV) radiation. All these make a contribution to more than 100 diseases and injuries. The report states that 8.2 million are deaths related to non-communicable diseases which are attributable to air pollution such as second hand exposure to tobacco. On communicable diseases such as cardiovascular diseases, cancers, and chronic respiratory diseases contribute to nearly two thirds of the total deaths that occur as a result of environments that are unhealthy (Taylor, 2005).

At the same time infectious diseases such as diarrhea and malaria which are often than not linked to poor access to clean water, sanitation, and management of waste have experienced a decline due to the improvement made in this areas. Environmental risks were also reported to take their greatest toll on young children and the elderly that is children under 5 years and adults between fifty and seventy-five years. This predicament can be prevented by environmental management. The health of the elderly is impacted on by non-communicable diseases why the health of fewer than five children is impacted on by respiratory infections and diarrheal diseases. According to regional statistics, the World Health Organization report finds that low and middle income countries south of Asia and west pacific regions were found to have the largest disease burden related to the environment in the year 2012 which was a total of 7.3 million deaths mostly as a result of indoor and outdoor pollution. Annual deaths in Africa were 2.2 million, 847, 000 in the American region, and 854,000 in the east Mediterranean region. 1.4 million Deaths were recorded in the European region, 3.8 million in the East Asian region and 3.5 million in the pacific region. The report also showed that there are different measures that can be immediately taken in the reduction of the burden of disease attributable to environmental determinants. These measures include the promotion of safer household water storage and better measures related to hygiene, use of cleaner and safer fuels, safe, and regulated use of toxic substances for both household and industrial use and finally occupational safety and health measures.

Quoting the words of former Director General of the World Health Organization Dr. Margaret Chan, a healthy environment underpins a healthy population. She also encourages countries to make their environments healthy especially where people live and work. Unless this is done morbidity and mortality rates related to different environmental risks will still remain high (Taylor, 2005).

8.2. WHAT IS ENVIRONMENTAL HEALTH?

Environmental health is a public health branch that is concerned with different aspects of natural and the built environment which affects human health. According to the world health, organization the field of environmental health addresses external factors that are physical, chemical, and biological to humans and all related factors and how they impact on behavior. It includes assessing and controlling these environmental factors and their potential to affect human health. Its target is to prevent disease and create environments that support health. This description does not include behavior that is related to the environment as well as behavior related to the environment that is social or cultural together with genetics. The natural environment includes things like air, water, and soil and also features in our surroundings that are physical, chemical, biological, or social. The built or man-made environment is inclusive of physical structures where people live, interact or work such as offices, home, offices, factories, farms as well as systems within the community such as roads and transportation systems, land use practices and the management of waste. Consequences of alterations made by humans to the environment which is natural include air pollution, are also parts of the man-made environment. The social environment that is excluded includes factors like exercise and diet, socioeconomic status and other influences within the society that may affect health (Battersby, 2013) (Figure 8.1).

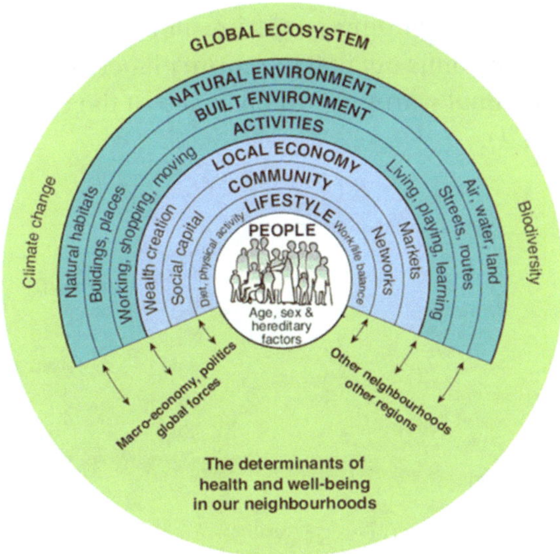

Figure 8.1: Determinants of health and well-being.

Source: Barton and Grant 2006 https://www.ehinz.ac.nz/indicators/overview/what-is-environmental-health/.

The World Health Organization also defines environmental health services as the services that are essential in the implementation of policies related to the environment through monitoring and control of activities. They also take the role of promotion of the improving parameters linked to the environment and encouraging the use of technologies and behaviors that are environmentally friendly at the same time healthy. They also play a leading role in the development and suggestion of novel policy areas.

Environmental health management mostly involves evaluating and examining the effects of man-made factors on human health, impacts of ecological systems and disease spread. Insulation of environmental and human health systems entail capturing the effects of human made and environmental health that are extended to public healthcare and the health of the public economy. It is essential to address and encompass the assessment and control of chemical, biological or physical agents of external origin and other related factor with behavioral impact that potentially affect health. The main target is creating an environment that is supportive (Taylor, 2005).

The global environmental health initiative which is spearheaded by the World Health Organization focuses on the improvement of environmental health of the community, prevention of disease epidemics, infant, and child mortality rates and in developing countries mortality rates. A recent report from the united nation points out that the entire planet might suffer drastically from the effects of global warming which is one of the environmental health problems (Figure 8.2).

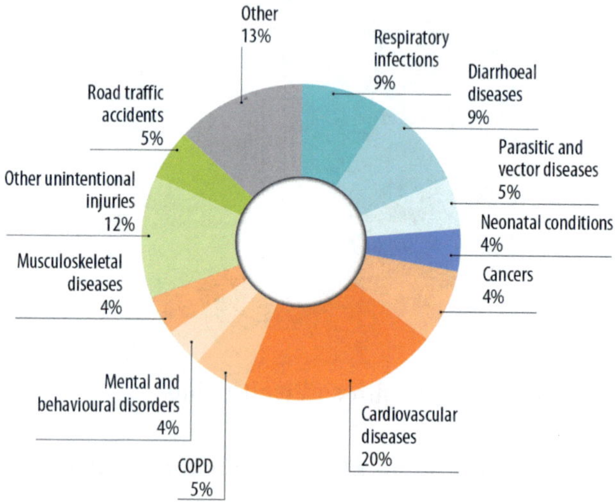

Figure 8.2: Main diseases contributing to the environmental burden of disease, all ages, world, 2012.

Source: https://apps.who.int/iris/bitstream/handle/10665/204585/9789241565196_eng.pdf;jsessionid=DD8170844AC547627332F6E09CEE30FA?sequence=1/WHO reports 2012.

8.3. DISCIPLINES OF ENVIRONMENTAL HEALTH

Environmental health can be divided into sub-disciplines which include environmental science, environmental, and occupational medicine, toxicology, and epidemiology. Other terms referred to in environmental health include environmental public health, public health protection and environmental health protection (Figure 8.3).

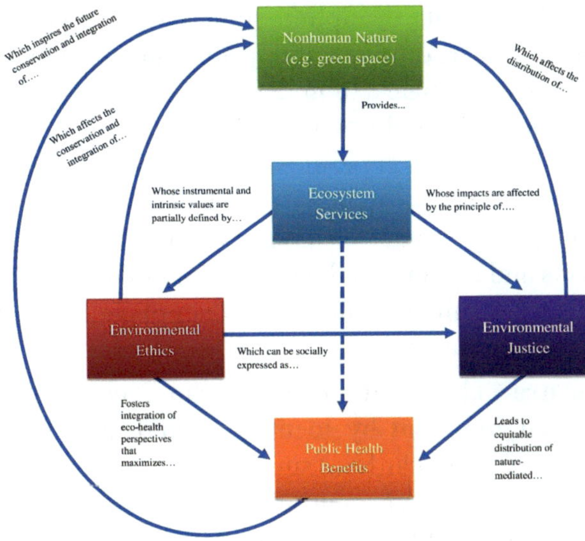

Figure 8.3: Viniece Jennings et al. "Finding common ground: Environmental ethics, social justice, and a sustainable path for nature-based health promotion," HealthCare doi: 10.3390/healthcare4030061, CC BY 4.0.

Source: https://commons.wikimedia.org/w/index.php?curid=75480372.

There are five basic disciplines that contribute to environmental health as mentioned above. Each of these disciplines makes different contributions by providing different information which is essential in the description of problems and solutions in environmental health some of which overlap (Gallant, 2008). The disciplines are as described below:

- **Environmental Epidemiology:** It studies the relationship that exists between exposures in the environment such as chemicals, radiation or microbiological agents and observational studies related to human health which observes exposure already experienced by people, those that are common in environmental epidemiology because it is not ethical to expose humans to agents that are known to or suspected to be the cause of a disease. Despite epidemiologists being unable to use experimental designs to test their hypothesis, they can directly observe environmental effects on the health of humans rather than estimating effects from animal studies.

- **Toxicology:** This discipline studies how exposures within the environment lead to specific health outcomes. This is generally in animals as a means of understanding possible health outcomes to humans. The field of toxicology has the advantage of being able to carry out experimental studies or randomized control trials because they can make use of animal subjects. Despite these advantages, there are differences in the biology of animals and humans and this brings the uncertainty in the interpretation of the results from animal studies and the implications they can have on the health of humans.
- **Exposure Science:** It studies human exposure to contaminants in the environment by identifying and quantifying exposures. Exposure science can be of support to environmental epidemiology by describing better environmental exposures that may lead to a particular health outcome, the identification of exposures that are common whose health outcomes may be better understood through a toxicology study, or it can be used in the assessment of risk to determine whether current levels of exposure might exceed levels that are recommended. The advantage of exposure science is that it has the ability to accurately quantify exposure to specific chemicals, but it fails to generate information about health outcomes like e environmental epidemiology or toxicology.
- **Environmental Engineering:** It applies the principles of both science and engineering in order to protect human populations from the effects of environmental effects that are adverse. These principles are also applied in the protection of different environments from potential deleterious effects of natural and human activities and the general improvement of the quality of the environment.

Environmental law-includes the network of treaties, statutes, and regulations, common, and customary laws which address the effects of activities by humans on the natural environment.

Information from environmental health principles such as epidemiology, toxicology, and exposure science can be brought together in conducting risk assessment for specific chemical mixtures, chemical that are specific or other risk factors to determine whether an exposure poses any significant risk to the health of humans. This means finding out whether an environmental exposure would likely result in the development of pollution-related diseases.

This information can then be used in the development of environmental health policies, for example, they can be used in the regulation of health policies, or it imposes standards of proper sanitation. Actions of engineering and law can be combined to provide risk management to minimize, monitor, and otherwise manage the exposure impact to protect the health of humans to foster the achievement of the objectives of environmental health policy (Gallant, 2008).

8.4. RELATIONSHIP BETWEEN THE ENVIRONMENT AND HUMAN HEALTH

Our environment means our physical surrounding and characteristics of the places where we live. I can also refer to the wider and natural world such as land, sea, and atmosphere. Human and environment interactions have existed since the beginning of time. For example, deforestation has been done by humans to create land for settlement, industrial, and agricultural use for centuries now thus leading to the alteration of the environment. Similarly, the environment has also gone to affect both humans and other biological organisms. The simplest example is how we change how we dress depending on the weather. The environment can influence human health both positively and negatively and vice versa (Gallant, 2008).

A good environment for example good climatic conditions, accessible clean water, fertile soil all of which are aspects of the physical environment enable people to live well and thrive. Environments that are harsh such as very hot climate, access to water and land that is infertile can make it make the survival of people difficult. Human beings are also affected by major environmental events such as earthquakes, floods, hurricanes; drought and volcano eruptions among others. Changes in environmental conditions can also affect human health such as global warming or exposure to hazardous agents or environmental contaminants. Environmental events can lead to the displacement of people; it can cause injury, loss of life and livelihood destruction. These events can lead to the damage of sources of water and water pipelines leading to contamination and spread of waterborne diseases. The relationship between human health and the environment changed with industrialization which began in the 18th century in the United Kingdom and it later spread to the rest of the world (Worthington, 2003)

Prior to industrialization the impacts of human activity on the environment was not notable since the technologies in use were not capable of modifying the environment on a scale that is large. At this time, the technologies applied were simple for example those practicing agriculture used hand tools for cultivation and these technologies had limited impact to the environment. Industrialization has allowed the exploitation of resources for examples when cutting down trees nowadays people use equipment such as chainsaws. In crop production, people employ the use of industrially produced chemical fertilizers and pesticides for crop production. These changes have overall and rapidly impacted on the environment. The link between human activities and the environment is complex and varied and can be grouped into two main types of activity. The first is the use of resources that are natural such as food, land, water, soil, plants, minerals, and animals. The second activity is waste production as a range of human activities such as agriculture, industry, and mining as well as wastes from our own bodies (Worthington, 2003) (Figure 8.4).

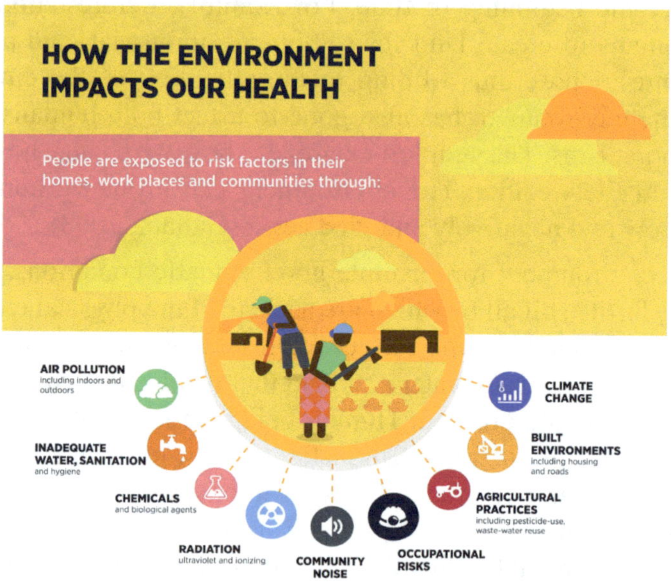

Figure 8.4: Impacts of the environment on health.

Source: https://dailyacts.org/faq-items/prevention-diseases/WorldHealthOrganization.

8.5. IMPORTANCE OF ENVIRONMENTAL HEALTH

We require clean water, air, plants, and a constant supply of food for our personal health and well-being. There is an interdependent relationship between humans and the environment. They both affect each other. Human actions and choices will affect the environment and similarly the health of the environment will affect the personal health of individuals as well as our communities, families, and economies (Taylor, 2005).

Maintaining a healthy environment is the key to increasing the quality of life and the life expectancy of the global community. Environmental factors are diverse and far reaching. Some include exposure to hazardous substances in the air, water, soil, and food, natural, and technological disasters, climate change, occupational hazards, and the built environment. Poor environmental quality has its greatest impact on those who are already at risk. Environmental health addresses societal and environmental factors that can increase the likelihood of exposure and disease.

We have learnt about the components or the sub disciplines of environmental health which are important in addressing environmental issues. Having basic knowledge on environmental health can be key and of tremendous significance to every individual in order to maintain an environment that is healthy. We deal with different issues related to the environment on a daily basis and having some knowledge will helping solve some of these issues (Worthington, 2003).

The different disciplines of environmental health are determined to identify all the aspects that could pose a threat to human health and solve some of the problems. The major driving factor is the concern about both human and environmental health. Therefore, there is a need to raise awareness about environmental health. Improvements in personal knowledge, skill, and practice can be responsible in the medication of individual behavior towards positive healthy practice for example in relation to good hygiene and sanitation practices.

Environmental health studies can be useful in informing policy especially towards championing an environmentally healthy and just world that is beneficial to all members of the community and the world at large. Environmental health is based on science and it entails the creation of scientific hypothesis in order to find answers. After a scientific hypothesis is made from a relationship let's say between the environmental health, the result can be used in policy formulation and planning for resources especially by the government. Credible hypothesis can be used for lobbying resources and direct them towards interventions related to the environment (Worthington, 2003).

The primary goal of environmental health is to prevent and promote the health of people. Environmental health is associated with the recognition, assessment, understanding, and the control of impacts of people on the environment and the impact the environment has on the public. The role of environmental scientists includes the following functions:

- Improving human health and their protection from hazards within the environment.
- Developing a like between the community and the local authority similarly between the local and higher administration levels.
- They can also act independently in the provision of advice on environmental health matters by designing and developing plans for environmental health actions.
- The enforcement of environmental legislation.
- They monitor and evaluate activities related to environmental health programs and projects.

8.6. ENVIRONMENTAL HEALTH PROBLEMS

Pollution and environmental concerns are interlinked and influence one another hence need to be addressed together. Key stakeholders along with community stakeholders should work together to reduce the environmental impact on public health.

- **Climate Change:** It is a result of greenhouse gases which trap heat from the sun warming the earth surface; increase in temperatures of the ocean which goes on to affect marine habitats and ecosystems; it also leads to global sea level rise which can account for land shrinking, massive floods and adverse weather events across the world.
- **Pollution:** It is one of the primary causes of environmental concerns related to change in the climate and biological adversity. Pollution includes all types such as air, water, soil, radioactive, noise, light, and thermal which affect the environment. Pollution and degradation of the environment paired with global warming has detrimental effects to the health of the public. Air and water pollution has been proven to also cause huge financial strain and

bring danger to both human and aquatic life. Implementation of law and stringent regulations along with educating people on pollution their causes and effects and how to prevent damage to valuable resources.

- **Deforestation:** It involves the destruction of trees and plants which provide resources such as oxygen, food, water, and medicine to the public population. The occurrence of natural wildfires, illegal logging, agricultural practice, and harvesting timber for commercial purposes, reduction of the supply of oxygen and an increase in the level of greenhouse gases. The level of biodiversity is one of the most important features of our natural ecosystems making up the environment which is danger of extinction due to global warming, pollution, and deforestation.
- **Diseases:** Mitigation and control are some of the major aspects affecting environmental health making radio logical, occupational, and industrial hygiene; drinking water that is safe; vector-borne control, diseases that are communicable and non-communicable which are necessary.
- **Hazardous Waste:** Management of waste that is hazardous, waste disposal systems, remediation of contaminated sites, prevention of leakage from underground storage tanks, emergency responses and the prevention of the release of hazardous materials into the environment has become essential. Improving systems of waste management can be essential in the prevention of cases of pollution in the natural resources which can trigger illnesses.

8.7. ENVIRONMENTAL HAZARDS

Humans come across different environmental hazards on a daily basis and this can be categorized into four to have a better understanding of environmental health:

- **Biological Hazards:** They are as a result of interaction or relations between organisms. Examples of biological hazards include bacteria, fungi, viruses, pathogenic microorganisms, disease vectors among others. When pathogens or diseases are transferred between two or more organisms, it is referred to as an infectious disease. One reason why humans suffer from pathogens and diseases is that they are infested by other organisms, which is a natural process but at the same time can be hazardous.

- **Physical Hazards:** They are physical processes that occur in the environment naturally, for example natural disasters like volcano eruptions, hurricanes, earthquakes, blizzards, and tornadoes. Physical hazards are sometimes considered as secret but this is not always, so for example ultra violet radiation occurs in an open manner on a daily basis. Ultra violet radiation can be termed as hazardous since frequent or too much exposure can lead to the destruction of DNA and can trigger the development of diseases such as cataracts or cancer of the skin.
- **Chemical Hazards:** They occur in the environment in two ways either natural or as a result of human activities. Examples of chemical hazards that are naturally occurring include heavy metals such as mercury and lead. Human made chemical hazards include a lot of chemicals produced synthetically and these include chemicals like pesticides, plastics, and even disinfectants. Some organisms also produce chemicals that are natural and can have a hazardous impact on the environment and human health for example elements contained in peanut and dairy can trigger allergic reactions to humans.
- **Cultural Hazards:** They can also be referred to as social hazards. They originate from within the community, choices related to behavior, occupations, and socio economic status. Cultural hazards include smoking of cigarettes which can be detrimental to the health of humans. Cigarette smoking is a behavioral and in people settle or live in the environment where the criminal activities or rates are high then it can be considered as a hazard based on the locality of people. In equal measure our diet choices, physical activity levels and our means of transport can be of great impact to our health and the ecological system at large.

8.8. CRITICAL ISSUES RELATED TO ENVIRONMENTAL HEALTH

Environmental health is one of the largest fields in public health because of the wide number of environmental factors that can impact human health and the quality of life. It addresses issues within our natural environment for example the case of clean water or sanitation but the same time they can be a consequence of human activities which includes norms in the society. The healthy people 2020s environmental health objectives have highlighted six areas which explain the crucial areas in environmental health that can be relevant to community health. Environmental health can address all human related issues and they include the following:

- **Air Quality:** We need air for our survival and there for it is a non-negotiable natural resource. If we don't take care of the substances, we release to the environment then it can have significant impact on our health. Poor quality of air can be linked to cancer of the lungs and COPD. Air pollution can also be linked to low birth weight. A study that was published in 2005 was able to come to the conclusion that infants born to pregnant women with exposure to high levels of ozone during the second and third trimester in comparison to their unexposed peers, were more likely to give birth to infants with a low weight during birth. The same effect was seen in those women smoking during pregnancy. In countries like the United States, the clean air act 1970 tried to change all that. This was the first time the country took the responsibility for protecting the quality of air for their citizens. They did this by the regulation of emissions that can be harmful from sources like cars and industries. In 1990, the act was expanded to address acid rain and ozone layer depletion and its function. In 2011, in the prospective report the agency for environmental protection projected that the clean air act would prevent more than 230, 000 early deaths by this year (Gallant, 2008).
- **Water and Sanitation:** According to the centers for disease control and prevention, an estimated 780 million people worldwide lack access to drinking water that is safe and approximately 2.5 billion lack adequate accesses to sanitation services like clean latrines or bathrooms. This has a staggering impact as an estimated 2200 children die worldwide everyday as a result of diarrheal diseases which are linked to poor hygiene and sanitation. A simple act of filtering and chlorinating water systems can result to a decline of common gastrointestinal diseases significantly. Using estimation, for every 1 million dollars invested in technologies that improve the supply of clean water, a country can get back 23 million dollars which is associated with medical and societal cost savings. This is so because the supply of clean water is responsible for the bulk

in the decline in mortality rates linked to children.

- **Toxic Substances and Hazardous Waste:** Toxicology is important in environmental health. Many of the materials needed in the advanced of industries and technology can cause harm to human beings. The use of plastics or heavy metals can also be harmful to the environment and the health of human beings and animals alike. One of the most publicized examples is the flint water crisis that was brought to light in the year 2015. The water was loaded with lead. This led to widespread outrage and fear from families living around the environment. The long-term consequences of drinking water contaminated with lead leads to long term complications to both children and adults. Some of the complications include brain damage as of the case of flint. The most disadvantaged people were children from low income families as more than 40% of those who were affected were from a population of people who were below the line of poverty. This example is significant in explaining the fact that environmental issues have a record of affecting those people who are already or the most at risk (Gallant, 2008).
- **Homes and Communities:** We spend a lot of our time in our homes, work places, and school. It is important that these places should be safe with the least of hazards, as well as an environment that is conducive for a healthy life style. For example, in a hostile environment people will be afraid to go out and exercise if there is a lot of violence. In a place where roads are poorly maintained there will be a lot of accidents that happen leading to injury or the loss of lives. A subfield that is emerging in environmental health is that of access to food. In some neighborhoods, there is lack of grocery stores and people often rely on convenience stores this means they do not have access to good quality foods such as fruits and vegetables which is vital for a healthy diet. Families in such areas are unable to make healthy food choices and this exacerbates their existing health disparities especially for low income families and those from minority groups. Some of the measures created to offset such impact are the urging of people to plant community gardens where the residents can plant healthy and produce that is fresh. Other measures include improving access to public transportation to grocery stores that offer full services and also farmers markets. Another measure can be to

change zoning laws as an incentive to retailers to offer healthy food options.

- **Infrastructure and Surveillance:** Information is a primary piece in any strategy for public health purposes. By understanding the risks presents and where environmental health professionals can deploy their resources better to prevent and mitigate them. The activities include the investigation and response to diseases and this field is referred to as epidemiology. It also entails screening populations for hazards and the establishment of programs for surveillance. Activities related to surveillance involve either going out and looking for health concerns that are particular or seeking information from professionals in other or closely related fields. These fields may include professional fields like medicine or agriculture and they can alert environmental health agencies when they encounter the problems or the concerns. A perfect example is the action in the surveillance of mosquitoes and abatement activities. The programs test mosquitoes for certain aspects including if dangerous infections are present like Zika virus as well as monitoring populations to ensure that the control measures put in place are working. This information can be of help to health officials to know what to watch out for on where to best spray mosquitoes and give alerts to the public if mosquito borne illnesses are spreading in an area (Gallant, 2008).
- **Global Environmental Health:** Environmental health professionals are prepared for climates that are warmer and wetter and are more likely to exacerbate threats to the public health globally. Increase in temperatures. As temperatures arise, it is accompanied by an increase in survival rate of disease vectors such as mosquitoes that can live and survive in areas that were previously cold for their survival hence increasing the number of people that can be affected by the vectors leading to diseases like malaria, yellow fever, and dengue fever. With the rise in sea levels, coastal cities and island nations are at risk of experiencing floods which potentially leads to the destruction of homes, displacement, and sometimes the death of people. These results leads to people overcrowding in areas considered being safe and this can lead to the rampant spread of infectious diseases. As people begin to experience extreme weather events, health officials anticipate years with the occurrence of sequential

storms and floods which destroy homes, facilitate the spread of diseases. Protection of the health of the planet is important in improving and maintaining the health of the global population (Worthington, 2003).

Other environmental issues include food safety concerns, land use concerns which include planning for smart use of land, vector control, disaster preparedness, climate change, waste management, and disposal among others.

8.9. RISK ASSESSMENT IN ENVIRONMENTAL HEALTH

Risk assessment is a process that is systematic in the identification and description of health hazards and for the determination of the extent of exposure to humans. The result is assessment of the likelihood of the occurrence of risks to the health of humans in the event they are exposed to them. Risk assessment is used.

The assessment of health risks functions to provide support to risk management and the decision making process. This means that the process of risk assessment should be entirely on science and research data. There principles employed in the process of risk assessment. Risk assessments are done to determine whether a health hazard has the potential to result in an actual health risk at the exposure level identified. The nature of a health risk and the measures of risk management should be taken (Babar, 2007) (Figure 8.5).

Figure 8.5: Principles of risk assessment in environmental health.

Source: https://thl.fi/en/web/environmental-health/risk-assessment.

There are four stages in the risk assessment process:

- **Characterization of the Hazard:** In this stage, the hazard is identified. Some of the questions answered in this stage include the adverse health effects caused by the exposure agent, the lowest adverse effect caused by exposure and determination of the exposure response relationship for the critical effect.
- **The Second Stage is Exposure Assessment:** Some of the questions answered include the environmental sources of exposure, the medium of exposure, exposure extent, those vulnerable and susceptible to exposure agents.
- **Characterization of Health Risks:** The adverse effects and exposure are compared. Some of the questions to be answered and determined include if the hazardous levels of exposure are exceeded and if there exist a safety margin and if a health risk exists and what the health risk is like and those at risk.
- **Management of Risks:** Risks are assessed and it is determined whether they can be managed for example by preventing them.

8.10. THE MANAGEMENT OF ENVIRONMENTAL HEALTH RISKS

8.10.1. Water Quality

Water is essential to humans; it maintains and supports the life of all biological organisms. There are numerous water sources but not all of them are safe for drinking. The water we drink daily is referred to as portable water and this is safe for human consumption without creating a risk of health problems. Water is used for different purposes other than drinking for household, recreation, and even industrial use. Other than, this water is also essential for agricultural purposes for the healthy growth of crops (Figure 8.6).

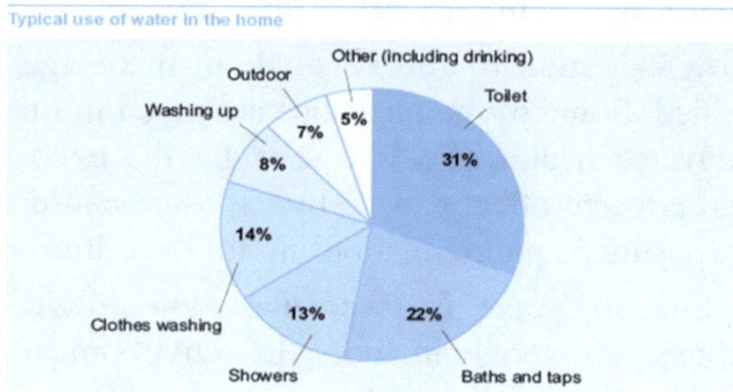

Figure 8.6: A chart showing the distribution of household water use.

Source: https://environmentalhealthxamk.wordpress.com/2017/02/11/week-6-managing-environmental-health-risks-potable-water/amp/.

There are two main sources of portable water. They include:

- Surface water which is collected on the ground or in a stream, river, lake, reservoir or oceans; and
- The second is ground water which comes from underground and can be obtained through well drilling. It is found in pores and spaces in between rocks.

Ground water and surface water are connected through the water or hydrological cycle; therefore, one can contaminate each other. Water can be contaminated by four major types of hazards which were discussed previously (Figure 8.7).

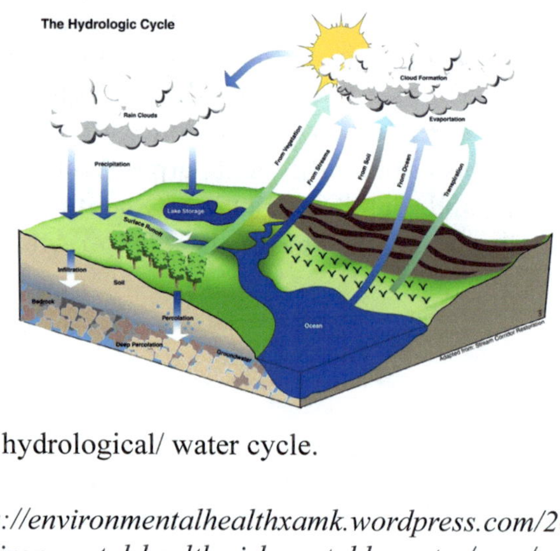

Figure 8.7: The hydrological/ water cycle.

Source: https://environmentalhealthxamk.wordpress.com/2017/02/11/week-6-managing-environmental-health-risks-potable-water/amp/.

Physical contaminants have an impact on physical appearance of water. Some of the physical contaminants include soil sediments or organic material that can be suspended in water sources and can be as a result of soil erosion.

Chemical water contaminants can be present naturally or from manmade activities. Some of the contaminants include heavy metals, nitrogen, bleach, salts, pesticides, and bacterial toxins; human and animal drugs.

Biological water contaminants are organisms present in water. They can also be called microbial contaminants some of the contaminants include viruses, bacteria, protozoa, and parasites.

Radiological water contaminants result in unstable atoms that can emit ionizing radiation. Some of the contaminants include plutonium and uranium.

Water quality is assessed based on different indicators such as alkalinity, water color, PH, odor, and taste, dissolved metals and salts, microorganisms such as coliforms, dissolved heavy metals, dissolved organics, radon, pharmaceuticals, and hormone analogs.

To convert water into portable water, a water treatment system is required. The water treatment must be standardized and have qualities that are satisfactory in order to produce drinking water that is not harmful to human health. Treatment involves removing contaminants to purify water for human consumption. Potential hazards are removed from water. Physical processes are used such as filtration and settling. Chemical processes used include coagulation and disinfection. Slow sand filtration is a biological process used. Apart from the measures used to make water portable, the conveying and distribution methods are also important. It is therefore important to keep residual disinfectants in treated water to treat pollution by bacteria that can occur in the process of distribution (Ricci, 2006).

There are seven steps followed in monitoring the quality of water. The first step is designing a program after which a monitoring system is selected. The next steps include selecting locations for monitoring and parameters for monitoring. The next steps include the decision on sampling frequency and duration; data management and use; other factors include quality assurance, quality control and safety (Ricci, 2006).

Implementation of several actions can also improve and maintain the quality of water. They include:

- Providing information on water quality to the public.
- Development of environmental education programs which can

be helpful to the community on how their actions can affect the quality of water.

- Supporting total catchment management programs as well as the provision of financial resources to communities.
- Developing programs to reduce pollution and also regulating activities that are industrial as well as the control of diffuse sources to prevent the pollution of water.
- Working with communities to deal with difficult problems related to water quality such as storm water pollution and urban runoff.
- Working with other government agencies, water boards and the local council in the implementation of water effective management strategies related to water quality.

8.10.2. Air Quality

- **Ambient Air:** Air is an important factor relating to environmental health. We are surrounded by air everywhere which can affect us significantly. Air pollution has become a problem with the world becoming more industrialized and modernized. It is therefore important to come up with a good management system to deal with the issue of air pollution. According to the world Health Organization, air pollution is the contamination of the indoor or outdoor environment by any physical, biological or chemical agents which takes part in the modification of the natural characteristics of the atmosphere (Figure 8.8).

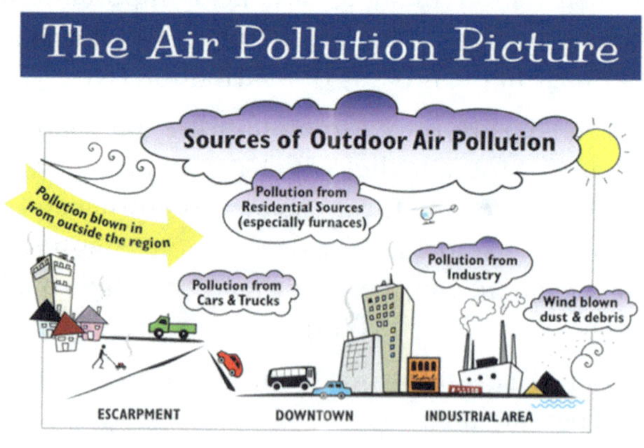

Figure 8.8: Air pollution in a picture.Source: eco-pollution.blogspot.com/2017/01/air-pollution-linked-with-eye-infections.html.

Pollutants of ambient air include the combustion of hydrocarbon fuels which produce air pollutants such as nitrogen oxides, hydrocarbons that are gaseous and carbon (II) oxide as well as other particulates such as lead.

Burning of fossil fuels such as oil and coal in generating stations, factories, and massive incineration of garbage. It produces ash, soot, and other particulates along with large quantities sulfur oxides which cause the rusting of iron, building damage, and nylon decomposition.

Insecticides, herbicides, radioactive fallout, dust from fertilizers, mining operations, and livestock feedlots. Sources that is natural such as volcanoes, biological decay, and forest fires.

There are numerous chemical contaminants in ambient air, some of which are noticeable such as Sulphur IV oxide and TRS, nitrogen IV oxide and other nitrogen compounds, PM10 and PM2.5 and O3 and CO (Figure 8.9).

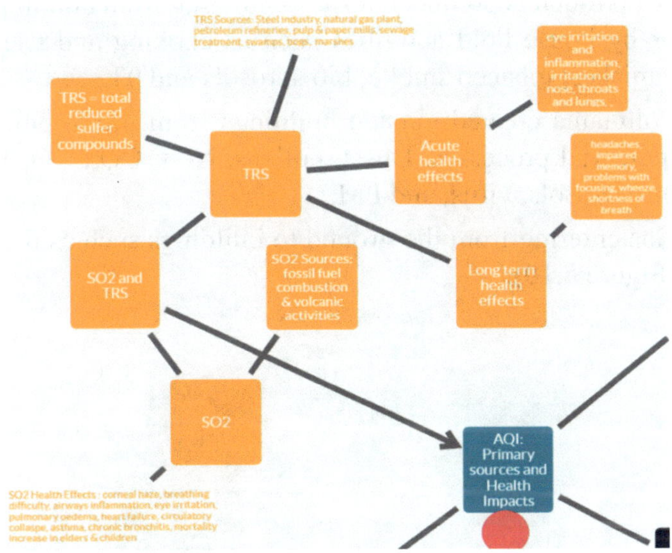

Figure 8.9: Air pollutants, their sources, and health effects.

Source: https://environmentalhealthxamk.wordpress.com/2017/03/18/week-11-managing-environmental-health-risks-ambient-air-pollution-and-noise/.

Some measures and solutions can be implemented in order to reduce air pollution and its effects. The government should encourage the use of public transportation. It is also important to practice energy conservation and the emphasis on the use of clean energy sources such as geothermal energy, solar, and wind. Another solution would be to develop the concept of reduce, reuse, and recycle and also raise awareness via campaigns and advocacy.

- **Indoor Air Quality:** Indoor air quality is related to air within and around structures and buildings since it is related to the health and comfort of inhabitants. According to the World Health Organization, approximately 4.3 million people die as a result of exposure to air pollution indoors. In addition to inhaling oxygen for survival, we also inhale dust, smoke, chemicals, microorganisms, and other pollutants contained in air. Air quality can affect the health of individuals and national productivity. Although people spend a lot of time indoors they are sometimes not aware of the health hazards present in the air. Indoor air pollution is dependent on an array of internal and external factors. This consists of:
- Air pollutants produced inside a building from building materials or by house hold activities such as cooking and use of paints, vanishes; tobacco smoke, bio aerosols and 03.
- Pollutants created outside buildings from traffic emissions and industrial processes. Dusts, pollens, NOs, CO_2, CO, VOCs, O3, bio aerosols, virus, and PM.
- Gas entering from the ground to buildings such as the radon gas (Figure 8.10).

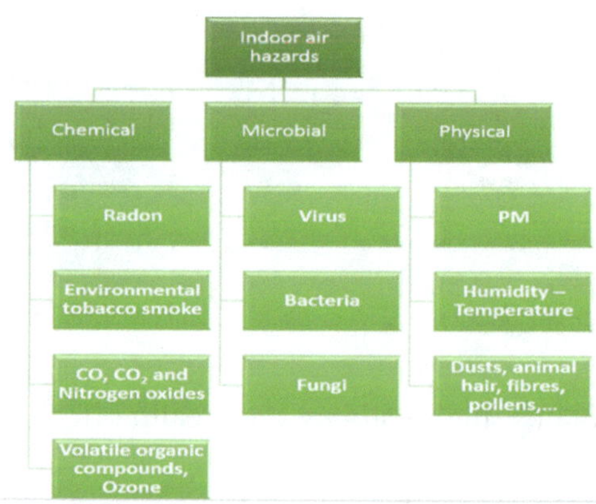

Figure 8.10: Categories of indoor air hazards and their examples.

Source: https://environmentalhealthxamk.wordpress.com/2017/04/02/week-12-managing-environmental-health-risks-indoor-air-quality-and-housing/amp/.

In order to control health risks that result from indoor air pollution, people should concentrate on pollutant sources or odors. They should also concentrate on the design, maintenance, and operation of ventilation systems; moisture and humidity as well as the perception of inhabitants and susceptibilities.

Good management of indoor air quality pays attention to a buildings system of heating, ventilation, and air conditioning; the layout and design of the building's space; outdoor air quality and supply; maintenance of equipment and the control of pollutant pathways.

8.10.3. Noise Pollution

Noise can be defined as sound that is especially loud, harsh or the confused kind. It can also be referred to as non-harmonious or a discordant group of sounds. There are different sources of noise pollution. At house hold level, they can include house hold electronics, social events, commercial and industrial activities and finally transportation (Figure 8.11).

Figure 8.11: Animated noise pollution posters.

Source: https://dailypost.in/news/miscellaneous/noise-pollution-us-threatens-protected-wildlife-areas-washington/.

Noise pollution has severe adverse health effects; excessive noise can disturb normal human activities. It can also lead to sleep disturbance, cause cardiovascular and psychological effects, cause performance reduction, and provoke responses of annoyance. It can also lead to changes in social behavior. According to the world health organization, traffic noise solely can harmful to the health of individuals (Ricci, 2006) (Figure 8.12).

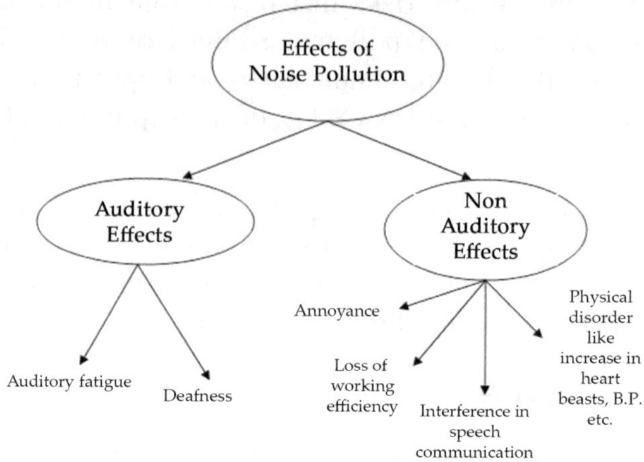

Figure 8.12: Effects of noise pollution.

Source: https://vijayshankarsaini.wordpress.com/category/environment-pollution/.

It is therefore necessary to develop a noise risk management system. It will also be essential in the removal and reduction of noise pollution sources and exposure:

- Removal and minimization of sources-establishment of permit requirement and determining permissible levels of sounds for example from vehicles and machinery.
- Exposure removal and minimization-planning of land use, carrying out an environmental impact assessment (EIA) and environmental permit.

An example is Europe which has a directive for the assessment and management of environmental noise. It is the main instrument in the identification of levels of noise pollution and to trigger action necessary for member states and the European Union level. It focuses on three key action areas.

- Determining exposure to environmental noise.
- Making public information on environmental noise and its effects.
- Prevention and reduction of noise in the environment where necessary and preservation of quality environmental noise where it is good.

It is applicable to noise to which humans are exposed to for example in built areas and public places or other quiet areas in open country, near schools, hospitals, and other places that are sensitive to noise. The directive does not apply to noise caused by the exposed person, noise from domestic activities, noise created by neighbors, noise at places of work, noise inside a transport means and those by military activities. The directive also requires member countries to prepare and publish noise management action plans and noise maps every five years. It is prepared for:

- Agglomerations which have more than 100, 000 inhabitants.
- Major roads those with greater than 3,000,000 vehicles a year.
- Major railways with greater than 30 train a year.
- Major airports.

When developing such actions plans for noise management, authorities are required to seek public opinion and approval.

8.10.4. Food Safety Management

Food is a source of vitality and energy for humans. It is important for growth, recovery, and strengthening the immunity. It is important to consume food that is of good quality and quantity. This means we have to include a variety of nutrients on a daily basis. Despite food being useful, it can also be a source of hazards. There are three types of hazards that can be exposed to food during harvesting, processing, transport, preparation, and storage. Sources can be either from chemical, physical, and microbial (Battersby, 2013). Chemical hazards can be of three types they include chemicals occurring naturally band are needed during harvesting, additives such as colorants, stabilizers, sweeteners, and preservatives. The final group is unintentional chemicals such as residues from antibiotics, agricultural chemicals and bio accumulated heavy metals. Toxic metals such as copper zinc, lead, and cadmium in their interaction with food product can create poisonous substances for example gray enamel ware plated with cadmium in its interaction with acidic products can be poisonous. Food additives are used to improve food quality for example by increasing freshness of food. Some of these additives are prohibited by law such as sulfites and it is required that they are added to food labels as ingredients.

Biological hazards pose the greatest threat to food safety which poses even a greater threat to the consumers. Microorganisms that infest food include fungi and bacteria. Bacteria including salmonella, *Listeria monocytogenes* and campylobacter can cause food borne illnesses similarly to fungi and viruses.

Unlike bacteria, fungi can grow in highly acidic foods with reduced moisture. Mold can produce spores which survive in harsh environmental conditions and grow when conditions are more suitable. Some of the molds cause allergic reactions while others produce mycotoxins than cause poisoning and intoxication. Other poisonous products from fungi include aflatoxins. At the same time microorganisms can also be beneficial in the food production industry, for example, bifido bacteria and lactobacillus are important in the health industry, some are also used in milk fermentation (Battersby, 2013).

Physical hazards can be naturally occurring or unintentional they include metal, glass fragments, plastics, stones, wood, and natural components.

In order to carryout proper risk assessment, certain procedures need to be followed. It may be accomplished by applying the HACCP process which can be applied in the food processing food chain in the identification of hazards and their control. It involves the application of the seven HACCP principles. The first principle is to identify the hazards. The second step is to determine Critical Control Points. The third step is the establishment of critical limits to ensure the CCPs are being met. The fourth step is the establishment of a system to monitor the control of critical control points. The fifth step is to establish corrective action when critical control points are not under control. The sixth step is to establish procedures for verification for confirmation of the efficiency of the system put in place. The last step is to document all procedures and records appropriate to the above principles (Battersby, 2013) (Figure 8.13).

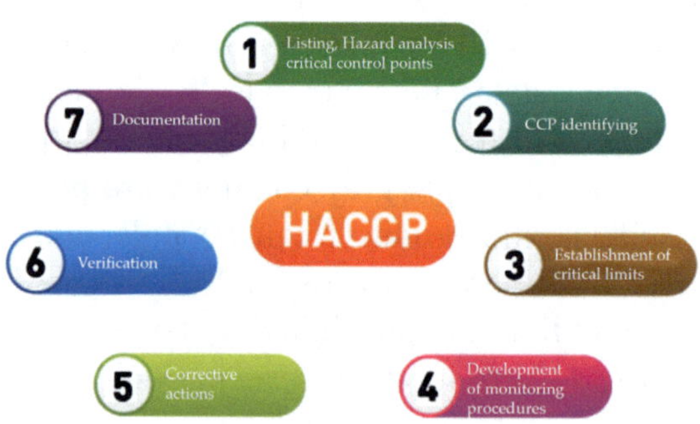

Figure 8.13: The seven HACCP principles.

Source: https://www.inecta.com/blog/what-is-haccp.

Another approach is the use of the rapid alert system for food and feed (RASFF). It is applied by members of the European Union. This approach allows it members to share information in a manner that is efficient and for example, it makes it easy to recall questionable products from the market. The European Food Safety Authority is involved as a member. Its notifications are classified by the commission according to how serious the risk identified is and its market distribution (Figure 8.14).

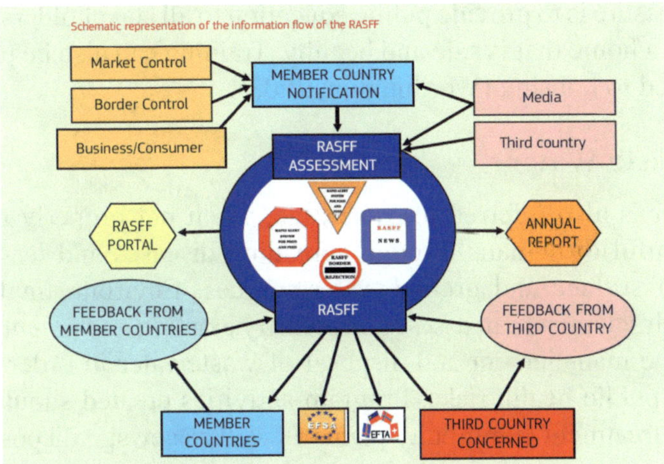

Figure 8.14: A representation of the flow information of the RASFF.

Source: https://paulineratzinger.wordpress.com/2017/03/31/week-8-and-10-managing-environmental-health-risks-food/amp/.

The Oxford Biomedical Research Food Science Division assesses the quality of food with biomarkers of food quality and freshness. Some of the biomarkers used are oil polarity, TBARs, total antioxidant capacity, and histamine levels.

8.10.5. Health and Housing

A healthy home means community inhabiting a home has the physical and social conditions are necessary for health, safety, comfort, and hygiene. Environmental health services can help to address public health issues related to housing and health.

8.10.5.1. Health and Housing Environmental Health Assessment

Provision of health inspections at home upon request. The public health inspections may include health and safety hazard evaluation and this includes indoor air quality, contaminants, control of pests, supply of water, solid, and liquid waste disposal, general safety, concerns related to structures and overcrowding. Building plans can also be reviewed from a public health perspective for new house developments and renovations of houses or buildings. Professionals can also provide guidance and recommendations to local authorities and occupants related to all stages of housing and development such as design, construction, occupancy, and demolition. Another measure is to provide public education to all stakeholders about how to maintain a home that is safe and healthy. Training can also be provided on issues related to housing (Worthington, 2003).

8.10.6. Waste Water

Waste water is also referred to as sewage, when not properly managed it can be harmful to humans through spreading diseases and leading to the pollution of surface and ground water sources. Environmental health or public health scientists can assess and identify existing and potential hazards related to the management and disposal of wastewater in order to prevent and reduce public health risks. Program activities created should focus on wastewater treatment community plants as well as sewage disposal systems that are on-site.

The process of assessment includes:

- The review of plans for on-site sewage disposal systems that are new and repairs from a perspective of public health.
- Provision of advice, guidance, and recommendations that are associated with on-site disposal of sewage, this is inclusive of information on appropriate decommissioning sites, if necessary
- Conducting inspections that are on site for installations that are new and the repair of disposal systems.
- Working with and approving installation of septic systems which are installed by on-site sewage professionals who are registered if applicable.
- Public health professionals can respond to complaints by providing inspections of existing sewage disposal systems where appropriate.
- Inspection of treatment plants that handle wastewater if there is a public health concern.

- Provision of advice, guidance, and recommendations associated with wastewater treatment plants.
- Provision of public education to home occupants on how to properly maintain on-site sewage disposal system and also reduce the risks that are associated with the discharge of sewage.

8.10.7. Solid Waste Disposal

Solid waste can become a public health concern if it is not well managed and disposed. Sites of waste disposal can attract animals and pests that spread diseases and leach pollutants that lead to the contamination of air, soil, and water. Solid waste can lead to the contamination of air if fires occur at these sites and such occurrence can lead to illnesses such as asthma. Authorities can work with communities and waste collection agencies to help prevent and control public and environmental risks to health posed by collection, storage, and disposal of solid waste (Ricci, 2006).

Measures to deal with the issue of solid waste management include the provision of inspection services for disposal sites and transfer stations. Evaluation of methods of solid waste collection, operation, and waste containment; the nature of disposed waste; control of pests; conditions of soil; conditions of ground water and leachate analysis. Reviewing plans for new or upgraded disposal site practices from a perspective of public health. Provision of advice, guidance, and recommendation to all stakeholders on issues related to collection, storage, and disposal of solid waste. Reviewing and provision of recommendations for safe decommissioning of disposal sites (Worthington, 2003).

Other measures include public education on best management practices on solid waste management and referrals on practices such as reducing, reusing, and recycling of solid waste together with the disposal of hazardous waste. This includes safe storage, collection, and disposal.

8.10.8. Emergency Preparedness and Response

Countries globally need to prepare for and respond to emergencies such as floods, forest fires, chemical spills, storms, food contamination and contamination of water supplies; outbreak of diseases. Environmental health services ensure public health considerations are included in the planning for emergencies and response activities (Smith, 2003).

Environmental health information provides advice, guidance, and recommendations to key stakeholders on public health issues related to emergency preparedness and response. Scientists can also take part in the development plans for emergency preparedness and response.

In the event of an emergency, assessment of emergency locations is done and advice is given to emergency response teams to reduce environmental health risks. Provision of temporary health inspections for accommodation, drinking water, food services, disposal systems for solid waste and wastewater. Provision of training to food handlers, sampling of drinking water and other emergency environmental health services (Smith, 2003).

9

Worst Environmental Disasters in Developing Countries: A Case Study in Africa

CONTENTS

9.1. INTRODUCTION

These are also called ecological disasters. An environmental disaster is defined as a catastrophic occurrence that is related to the environment as a result of human activity. This can be easily mistaken as a natural disaster which is different from an environmental disaster in the sense that it is an outcome of natural occurring catastrophes that are related to nature more than human activity. In the case of environmental disasters, man interferes with the functioning of the eco-system. An eco-system is defined as an existing community of living organisms as well as the non-living components of the environment and creates harmony as a system as well as develop an interdependence on each other. Biotic and Abiotic components are the two main components of an eco-system linked together through nutrient cycles and energy flows. Interference with the eco-system creates long lasting effects hammered on the environment that later on grow to vulnerability of humanity to disasters. These lay in an array of categories. We have the death of animals, plants, human life disruption, and a possibility of migration. Some of the examples of environmental disasters are; landslides, floods, wildfires, earthquakes, drought, tornadoes, tsunamis, volcano eruptions, cyclones, as well as large insect invasions.

Environmental disasters affect vital areas of the human day to day life like agriculture, the economy, human health, and biodiversity. This rapidly increases climate change. The need to feed the rapidly increasing population, the need to satisfy production demands and many other thresholds we need to meet for the sake of humanities sustainability, places us at the receiving end of the environmental disasters. More than often in the current times, we live in a time where greed and the need to make more money dominate the order of business which costs us greatly. This chapter seeks to bring forth the results and effects of environmental disasters in the continent of Africa. This is classified as the developing countries or the third world countries. We will look at various disasters and the effects caused as well as are careful to include the fact that these countries are disadvantaged in one way or another. Poor development, lack of ideologic liberations, political liberation, as well as battling with matters such as poverty and corruption in the 21st century is some of the featured facts in this chapter. The culture and 'backward' nature of the people that inhibit Africa, the level of poverty and immense population numbers, lack of education and lack of health facilities to deal with these kids of emergencies are some to the factors that will be featured in this document (Simelane and Thokozani, 2019).

9.2. DEFORESTATION IN AFRICA

Deforestation is a term that refers to the permanent removal or uprooting or a tree to make room for something that is not a forest. This in many cases is to make room for agricultural work, buildings, roads, go-downs, and industrial work, the usage of timber for fuel or for mining or drilling into the earth's crust for resources (Figures 9.1 and 9.2).

Figure 9.1: A forest cleared by human logging.

Source: https://storymaps.arcgis.com/stories/d17484098e17420f897881d9e3a98a8a.

Figure 9.2: Experts estimate that a chunk of forest the size of a soccer field is lost every second to deforestation.

Source: https://www.livescience.com/27692-deforestation.html.

According to World Wildlife Fund, forest cover is above 30% of the earth's land surface. These forests provide a living for the African people. Food, water, traditional, and natural medication, as well as fuel for the, millions of people that rely on it. The forest offers a huge fraction of labor and work for these same people. Forests can be classified as a resource, though some of them lay as large, underdeveloped swaths of land that can be converted for the sake for agriculture or grazing. Today, deforestation is a worldwide embraced practice. According to the World Bank, an estimated 3.9 million square miles of forest disappeared at the beginning of the 20th century. In the last 25 years, forests have shrunk down in size by an estimated 502,000 square miles which is an area larger than the country of South Africa (Simelane and Thokozani, 2019).

The main causes of deforestation in Africa are agricultural expansions, timber extraction for fuel, cattle breeding, oil extraction, mining, dam constriction, infrastructure, human settlement, military conflict, road construction, and firewood collection among other reasons. Forests can be located in the tropics to high-latitude areas. They feed majority of the microbes in the eco-system, therefore making them a very important part of man's existence.

The loss of trees and other types of vegetation in such large amounts on a large scale has an effect on climate change through the environment. Soil erosion, flooding, increased greenhouse effect, desertification, lack of rain and low water quality is some of the effects that people in these places face due to deforestation. Some of these effects are caused by the basic lifestyle and livelihood for people on a daily basis. Forests offer basic provisions like firewood for cooking in a homestead, for example in countries similar to Uganda, there is a huge reliance on firewood, timber, and charcoal for fuel to cook and run basic household activities. Over the last 25 years, the country of Uganda has accumulatively lost about 63% of forest cover due to family and homestead support as well as construction and grazing. The loss of trees in this rapid manner also leaves us with large amount of carbon dioxide to deal with. Naturally, trees absorb carbon dioxide and release oxygen which humans and animals consume. This leaves us with no air purification method. Across Central Africa, the Congo Basin is home to about 80,000,000 people that depend on it for food, charcoal, and medicinal plants. This rain forest is also responsible for the rain and water patterns across many countries in the continent of Africa. The rate of deforestation in the basin is historically lower than that in the Amazon or in the Southeast Asia. Though, the rate is gradually and dangerously increasing and in particular, in the Democratic

Republic of Congo. DRC has over the years lost the second largest area of tropical primary forest compared to any country in the world as of 2018. This is considered a significant loss. By 2100, the primary forest of DRC is estimated to have been completely gone (Simelane and Thokozani, 2019).

This should not be looked at as just an issue of trees. Studies show that moisture regulation as well as rainfall regulation is intertwined with forests and more so, tropical rainforests. This is also closely related to temperature regulation in the atmosphere. The Congo Basin being the main source of rainfall in the Sahel region, the chances of air producing twice as much of rain as is typically the case when it passes over an area covered with extensive vegetation goes down significantly. Meanwhile, in the Horn of Africa and regions nearby are becoming increasingly dry with Somalia and Ethiopia experiencing severe drought that led to donation of millions in relief help in form of food and water. The continued deforestation opens doors to extreme weather conditions such as El Nino and souring temperatures affecting life in Africa spurring more and deeper crisis. With central Africa being home to some of the richest minerals and resources from the earth's crust such as wood palm oil, rubber, and others, there needs to be deeper studying in order for mankind to understand the importance of the Congo Basin. This is also a way to educate those living in these areas the huge importance of these kinds of forest cover have on the environment and how to rehabilitate the very brutally exploited forests of Africa (Meeme, 2020).

9.3. GULLY EROSION IN AFRICA

A gully is a landform, a ravine that has been formed by strong running water activity which erodes the soil sharply typically on hillsides. Gullies look very much alike to huge ditches or small valleys but are tens of meters deep and wide. Gully erosion is defined as the gully forming process. This process is more prone to the hillsides after the clearing of vegetation through deforestation or over-grazing mainly. The soil is eroded and carries away with ease by the water coming in with great strength and in large amounts in thunderstorms or by regular rainfall. The gully could grow in length meaning headward erosion means at a Knick point. This erosion is key in the result end of interflow and surface runoff. Gullies greatly reduce the productivity of farmland because they incise into the land and they also produce a lot of sediments that in the long run end up clogging water bodies (Meeme, 2020).

In Nigeria, the gully erosion crisis has been an ongoing crisis since before 1980. This has over the years affected the communities, large, and small, in many ways. The biggest result here is land degradation due to the serious damage to the earth's soil. It is also an environmental, ecological, and economical disaster because the livelihood of people is affected as well as future possibility of using the land to bring forth any use for the community. Loss of lives is also one major disadvantage that comes with this degradation. In Nigeria alone, there is an estimated 3,000 gullies and in southeast Nigeria, the exposure has expanded significantly from 1.33% in 1976 to a reported 3.7% in 2006. This makes the region the most affected region in the country (OCHA Services, 2019) (Figure 9.3).

Figure 9.3: A gully in Calabar, Nigeria.

Source: https://paradisenews.ng/39615/gully-erosion-threatens-residents.

The main causes of this gradual but deadly degradation are the surface runoff causing the gullies to form. With more water activity inclusive of rain, poor soil infiltration, unfavorable catchment shapes or poor drainage, the gullies grow wider and deeper. However, in the case of Nigeria, the causes can be mainly attributed to various activities:

- Poor road designs, construction, and abrupt termination of drains.
- Poor waste management: practices Such as dumping refuse into dreams and waterways can always lead to blockages and an eventual flooding situation.
- Unsustainable land use such as improper sand mining always strips land of its topsoil and tilling is associated with crop cultivation that loosens the topsoil creating an immense possibility of erosion by a path for runoff.

Gully erosion is mostly concentrated in several towns and states in the southeast parts of this Nigeria. The crisis continues to affect all Nigerians indirectly for example homes and structures regularly collapse as the galleys continued to expand eventually transforming the region into a badland. The major impacts caused by gully erosion are:

- The loss of infrastructure mainly houses, roads, and other major real estate.
- Continuous loss of agricultural produce, agricultural land, and vegetation.
- Increased siltation levels in rivers causing an increased possibility of flooding.
- High levels of sedimentation in waterways.

This leaves a lot of room for land degradation and a great deal of environmental insecurity in Nigeria (Figures 9.4 and 9.5).

Figure 9.4: Gully erosion destroys houses in Anambra community, Nigeria.

Source: https://www.independent.ng/gully-erosion-destroys-houses-in-anambra-communities/.

Figure 9.5: The Umuchiana ravine in Nigeria.

Source: https://www.climatechangenews.com/2020/01/20/erosion-crisis-swallows-homes-livelihoods-nigeria/.

The Estimated damages of the economy brought about by gullies mainly in southeast Nigeria could be up to $1,000,000 yearly in comparison to the agricultural yield counting losses of 30% to 90% around the areas affected. The same battle being fought by Nigeria with soil erosion can be seen to translate in other parts of the African continent. In Sub-Saharan Africa, soil erosion accounts for 77% of land degradation and about 22% of the land can be used but it is also greatly threatened. Uganda faces the same battle having 70% of the land being degraded and also the devastated grazing areas of the Masai landscapes in Tanzania also have been costed greatly by overgrazing. In recent events, we have seen gullies and other environmental disasters being politicized in political campaigns leading to a great deal of political greed, misinformation of the public as well as major corruption in Africa. Gully erosion can be prevented but good strategic planning. Management and maintenance of remnant vegetation along drainage lines and banning grazing in these areas, increasing water usage in the planting of deep-rooted trees and vegetation, and carefully managing drainage and waste disposal channels can contribute miles into the restoration of the soil and preventing further damage in the valuable future (Simelane and Thokozani, 2019).

9.4. EARTHQUAKES IN AFRICA

An earthquake is defined as the shacking of the surface of the earth which results to a sudden release of energy to the earth's lithosphere creating seismic waves. Earthquakes range in different sizes and from those there are weak earthquakes that cannot be felt and there are also very strong and violent earthquakes usually enough to propel objects and people into the air in the process causing destruction in the affected areas. The seismicity or the seismic activity is the frequency, type, and size open earthquake experienced in a given area over a period of time. The word tremor is also used for earthquakes but those that do not have seismic rumbling. On the earth's surface, earthquakes manifest themselves by disruption on the ground and also displacing items as well as infrastructure. At times the epicenter of an earthquake is large and located offshore, the seabed may experience sufficient displacement which causes a tsunami. Earthquakes also related to triggering landslides and occasionally bringing forth volcanic activity. Earthquakes can be caused by natural occurrences, but they can also be caused by human activity which generates seismic waves. Some of the human activity that translates to earthquakes is volcanic activity induced by man, mine blasts, and nuclear tests. Earthquakes occur more often in areas of geologic faults, narrow zones where rock masses are in motion in relation to one another. Major fault lines of the world are located at fringes that have huge tectonic plates making up the earth's crust (Simelane and Thokozani, 2019).

Very little was understood about earthquakes until the 20th century when the emergence of seismology began which is the study of earthquakes and how they occur and affect the earth and the environment which has later on yielded to answering many questions about earthquakes. Earthquakes oh come mainly in belts containing major tectonic plates. The most important earthquake belt is the Cricum Pacific belt that affects many populated coastal regions along the Pacific Ocean. A second belt known as the Alpide belt passes and dominates through The Mediterranean region throughout Asia and connects with the Cricum Pacific belt in the East Indies. A tectonic earthquake can occur anywhere in the earth as long as sufficient stored elastic strain energy is provided to drive fracture propagation throughout the fault plane. The slides of the fault plane move past each other smoothly only if there are no restriction along the fault lines which leads to a stick-slip behavior. After the fault lock, the continued relative movement between the plate's leads to high stress which is stored strain energy kept in volumes all around the fault plane. This will be a continuous process until the stress has risen significantly and is able to breakthrough, sliding over the locked portion of the fault plane and releases the stored energy. This energy is released and his combined with radiated elastic strain seismic waves, which are frictional and heating up of the fault surface create cracking of the rock, thus causing an earthquake. This process is a gradual buildup of the strain and the stress accumulated and punctuated by an occasion of sudden earthquake failure and is known as an elastic rebound theory (IFRC, 2011) (Figure 9.6).

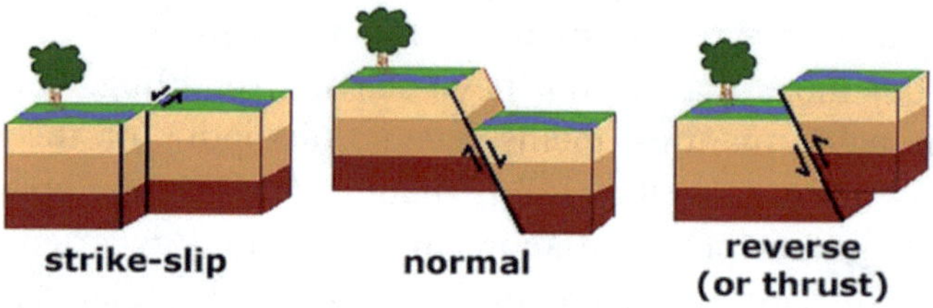

Figure 9.6: Three types of faults, strike-slip, normal, and reverse.

Source: https://www.geologypage.com/2017/10/three-main-types-faults.html.

There are three main types of faults. All of this may cause an inter-plate earthquake: normal, reverse, and strike-slip. Normal and reverse faulting are given as examples of dip-slip which is the displacement along the fault lines in the direction of the deep where movement on them involves a vertical component. Normal faults mostly occur in aerials where the earth's crust is short and such as divergent boundaries. Reverse faults also mainly occur in areas where the earth's crust is being expanded called convergent boundary. Reverse fault particularly those found along convergent plate boundaries can be associated with some of the most powerful earthquakes including almost of a magnitude eight or more. Earthquakes associated with normal faults are generally less in magnitude which is less than seven. Strike-slip fault are steep structures where the main two sides of the faults slip in a horizontal manner passing each other transforming boundaries particular to this type of fault. Many earthquakes are caused by movement or faults comprising of both dip-slip and strike-slip which is known as an oblique slip. For every Unit increase in magnitude there can be roughly a thirtyfold increase in release energy. For example, an earthquake of magnitude 6.0 can release 30 times more energy done an earthquake of magnitude 5.0. An earthquake of magnitude 7.0 releases up to 900 times more energy when that over 5.0 magnitude (IFRC, 2011) (Figure 9.7).

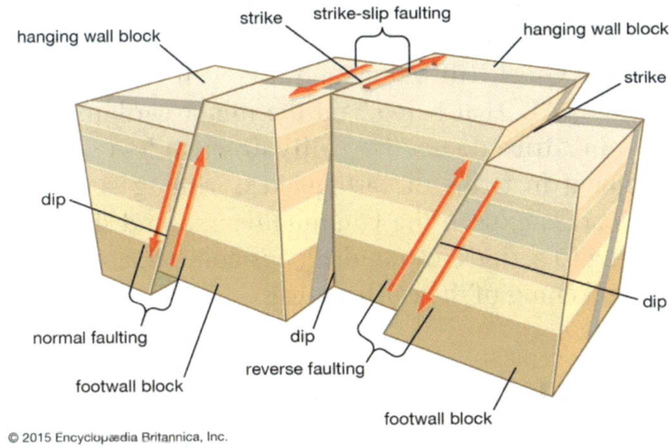

Figure 9.7: Types of faulting in tectonic earthquakes in normal and reverse faulting, rock masses slip vertically past each other. In strike-slip faulting, the rocks slip past each other horizontally.

Source: https://www.britannica.com/science/fault-geology.

In Africa, earthquakes have been experienced in several countries. Algeria, Democratic Republic of Congo, Botswana, Egypt, Eritrea, Libya, Guinea, Morocco, Malawi, Tanzania, Zambia, Zimbabwe, Rwanda, and South Africa have all experienced earthquakes. Majority of the seismic activity In Africa his concentrated mainly along the East African rift system with an addition of the north and East African regions as well as the Congo basin Seismic thickness may differ in correspondence with lithospheric thickness. In regions of thin lithosphere, the seismic thickness is limited to 20 kilometers, while regions with thick lithosphere, the seismic thickness are shoots up to 40 kilometers. In South Africa, there are several earthquakes that have happened leading to the loss of lives and the destruction of property. On 11th November 2013 an earthquake was felt as far as parts of north and South and Johannesburg which lasted a period of 10 to 15 seconds. The epicenter was the University of Johannesburg with a magnitude of 4.0. No lives were claimed, and minor property destruction was sustained. As a result of mining, on 9th March 2005 2 miners died while underground after the occurrence of an earthquake with a magnitude of 5.3 occurred (OCHA Services, 2019).

Studies have shown but Africa continues to experience fewer earthquakes due to the tectonic plates. Where you might see tectonic activity is likely to be in the Great Rift Valley and it is well African plates started to pull themselves apart. That being said the minor earthquakes that have been experienced In Africa do not disqualify or claim less vulnerability for the African continent in terms of earthquakes. Mining activities by man that contain explosives expose the communities around these places two earthquakes leading to serious economical, agricultural, and environmental degradation by weakening of the earth's crust.

9.5. CYCLONES IN AFRICA

A cyclone is a large skill air mass that rotates around a center of low atmospheric pressure. Cyclones can be characterized by an inward spiraling wind that continues to rotate around a zone of low pressure. The largest low-pressure systems in the world are polar vortices and extra tropical cyclones are of the largest scale. Warm-core cyclones, for example the tropical cyclones and the sub-tropical cyclones also lie within the largest scale of cyclones. Drawing energy from the surface of the sea and maintaining immense levels of strength as long as it stays over warm water, a tropical cyclone can generate winds exceeding 119 km per hour. In some extreme cases, the winds may reach up to 240 kilometers per hour and well surpass 320 km per hour. Such events contribute highly to the collection of water through the creation of a cyclone and a serious hazard is formed for coastal areas in tropical and subtropical areas of the world (MASIE, 2020) (Figure 9.8).

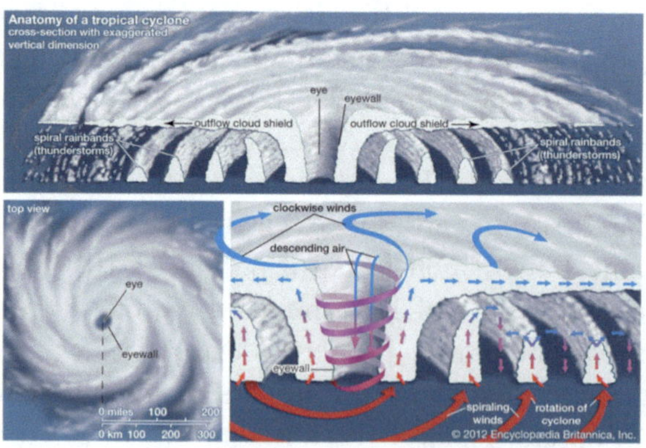

Figure 9.8: Image of the anatomy of a cyclone as well as a cross section.

Source: https://www.britannica.com/science/tropical-cyclone.

Around the world, cyclones are identified by different names. For example, in the North Atlantic Ocean and the eastern North Pacific, they are called hurricanes. In the Philippines, Japan, and China, they are identified as typhoons. In the western South Pacific as well as the Indian Ocean, they are referred to as cyclones, tropical cyclones, or simply just cyclones. There are two types of cyclones namely; mid latitude cyclones which are the main causes of winter storms in the middle latitude and tropical cyclones which are also called hurricanes. A cyclone goes through various stages as it grows into a mature tropical cyclone. The storm always begins as a tropical disturbance occurring when loosely organized cumulonimbus clouds start to show signs of a weak circulation. Reaching speeds of 36 km per hour, it is referred to as a tropical depression and at speeds of 63 km per hour; it is referred to as a tropical storm. When the storm reaches speeds of 119 km per hour, this is when it can be formally classified as a tropical cyclone (MASIE, 2020) (Figure 9.9).

Figure 9.9: A live tropical cyclone from an aerial view.

Source: https://www.shutterstock.com/video/clip-12976517-4k-eye-hurricane-space-perfect-for-eye-wall.

The main effects of tropical cyclones our heavy rain, extremely strong winds, tornadoes, and large storm surges near landfall. Depending on the size and the strength of the cyclone, the effects could vary. Location is also very key to the causes and the effects of cyclones. Cyclones remove forest canopy and change the landscape near the coastal regions through movement and reshaping of the sand dunes causing massive erosion along the coast. They can lead to great landslides and oxygen concentration in caves. There is also a lot of marine disruption in the event of a cyclone happening. After the cyclone has passed, there is serious devastation incurred by people. Loss of houses and homes, business structures and infrastructure, real estate, power lines, and communication technology is also disrupted greatly. Water contamination and the creation of standing water leaves opportunity to water-borne diseases that are contracted by people causing outbreaks in these affected areas (Meeme, 2020).

During the last two centuries, there have been a recorded number of 1.9 million people dead worldwide from cyclones. An estimated 10,000 people die each year from these cyclones and billions worth of property and goods go into ruins taking humanity years to recover from the effects. Coastal ridges, limestone cave stalagmites, and landscapes are classified as natural recourses that harbor great damages from cyclones effects (Figure 9.10).

Figure 9.10: An aerial view of Mozambique's Sofala Province shows standing water after a cyclone Idai dated March 14th, 2019.

Source: https://www.worldvision.org/disaster-relief-news-stories/2019-cyclone-idai-facts.

Cyclone Idai made landfall in Mozambique as a category 2 storm. The heavy rains and strong winds caused a lot of flash flooding and death of thousands of people along massive destruction or crops and property. A few weeks later, on April 25th, 2019, cyclone Kenneth hit hard on Northern Mozambique about 600 miles away from cyclone Idais impact zone. Catastrophic flash flooding from the storm claimed the lives of 2.2 million people in Mozambique, Malawi, and Zimbabwe and there were unexplainable and unfathomable amounts of damages caused. Hundreds and thousands of people were displaced and became homeless; others lost their family members and friends. These communities were devastated with the loss of their homes and livelihood. Death due to cholera, malaria, diarrhea, and respiratory diseases rose very high due to the poor standards and living conditions. With countries similar to these ones, there is lack of medical facilities and emergency teams that can deal with disasters of this magnitude. An already suffering population from unemployment, lack of food, lack of water and sanitation, drought, pests among many others could not handle this without driving themselves deeper into the devil that has continued to rob Africa of its potential; poverty. Disasters of this nature do take place in Africa and due to the opportunity created by poor roads, deforestation, and desertification as well as drought, the effects are only too great, and the vulnerability only continues to grow by the day (Meeme, 2020) (Figure 9.11).

Figure 9.11: Damage from a cyclone in South Africa.

Source: https://www.bloomberg.com/news/articles/2019-03-25/damage-from-southern-african-cyclone-may-top-1-billion-un-says.

9.6. DROUGHTS IN AFRICA

A drought can be identified as a prolonged period of time abnormally without or with low amounts of rainfall. This often leads to a great shortage of water. This can last up to a period of a few months or years. The effects on the eco-system and agriculture can be substantial regionally and can cause dangerous economic harm to the localities affected. Dry seasons in the tropics increase the chances of drought significantly and in rare events bushfires. Fast evaporation of water due to high temperatures can also greatly worsen the effects of drought. Drought denies the eco-system the water content it desperately needs for crop growth in order to sustain human life and their animals. In deserts like areas, the eco-systems have relatively low productivity. The worst and most prolonged drought recorder in history occurred in the Atacama Desert in Chile over a period of 400 years. Droughts affect humanity and their societies as it is considered a disaster; a supernatural or human caused. It is documented early in the lives of humans as form the Bible first in the Epic of Gilgamesh and tied to the story of Joseph, and it is later on covered in the Exodus story from Ancient Egypt (Simelane and Thokozani, 2019) (Figure 9.12).

Figure 9.12: Women who fled drought line up to receive food distributed at a camp for displaced persons in the Daynile neighborhood on the outskirts of Mogadishu, Somalia, May 18, 2019.

Source: https://www.hrw.org/news/2019/09/04/small-victory-somalis-un.

Hunter-gatherer migrations in Chile dated 9,500 BC are linked to this as well as the humans Out of Africa are linked to the exodus drought and it later on infiltrated the rest of the world 135,000 years ago. There is an existence of rain making rituals that are inclusive of dances and human sacrifice to prevent or avert drought. These practices were later on replaced by civilization and are currently known as water management.

Droughts can be categorized into three major types:

- Meteorological drought which occurs when there has been less than average precipitation for a long period of time.
- Agricultural drought which affect the crop production or the ecological range.
- Hydrological drought brought about by water reserves like lakes fall below a locally significant threshold. The receding levels of water are not noticed due to the slow and gradual use of the water and not enough to replenish it.

As the drought continues to persist, all the surrounding conditions continue to worsen, and the impact faced by the surrounding communities face serious effects. A drought can be caused by a lot of reasons. There are however causes that carry more weight than others. These are:

- Precipitation deficiency; this is any product of the condensation of water vapors in the atmosphere under gravity from clouds. The main forms are drizzles, ice, sleet, rain, snow, hail, and graupel. Major precipitation includes convective, stratiform, and orographic rainfall.
- Dry seasons; caused by low humidity with watering holes and rivers drying up.
- Erosion and human activity; weakening the topsoil that is rich and important for crop productivity but carried away by strong winds and waters from rain.
- Climate changes; global warming being the major increasing world rainfall. Climatic change activities will trigger the occurrence of drought with a substantial impact on the agricultural system and this causes way worse effects in the developing nations that have very high populations in communities.

The consequences of a drought can be divided into three groups; social, environmental, and economical. The environment faces serious degradation with the dropping of water levels and low rainfall levels, the drying of wetlands, loss of biodiversity, the deteriorating health of trees, and the appearance of pests and dendroid diseases. Economic losses are low agricultural produce, low fishing and game produce, high cost of food and production, low energy production, loss of tourism and transport revenue, etc., social costs include a negative health response from people, possible water rationing, increased population, high food-costs, failed harvests induced stress as well as poor living conditions. Droughts cause poor water quality due to low water flow causing pollution and increased contamination of the water. The effects attached directly to this are:

- Low crop yields;
- Dust bowls causing further erosion of the soil;
- Famine;
- Hunger;
- Mass migration;
- Malnutrition, dehydration, and related diseases;
- Social unrest;
- Wars among men over the remaining resources and most of them being natural resources;
- Cyanotoxin accumulation in food chains and water supply which lead to cancer and an eventual low expectancy rate.

In Africa, drought is one of the most seriously faced challenges that continues to eat into the lives of the African communities and ties them down to poverty and poor standards of living. There have been several campaigns to save Africa at large from this problem, the biggest one being “We are the World” campaign in 1985 by musicians from the united states of America aimed at raising money to help in relief food and water supply for those suffering. To date, lives are still being lost, crop produce goes to waste, animals die, and a continuous chain of loss creates an eerie silence and darkness in the lives for the communities that continue to chew on this tough bone (Figure 9.13).

Figure 9.13: Residents of Kaeris in Turkana North draw water from a shallow well.

Source: https://www.the-star.co.ke/news/big-read/2019-10-29-from-drought-hunger-to-floods-no-respite-for-turkana-residents/.

East Africa has been a primary focus of drought studies over the years. Countries around the Horn of Africa also suffer the same fate, struggling to survive the drought and extreme conditions in those areas. All across Ethiopia, Kenya, Somalia, and majority of the autonomous Somaliland, there are 10.7 million people facing hunger and starvation caused by drought (Figure 9.14).

Figure 9.14: Sheep and goats dead due to the progressing drought situation in Somaliland.

Source: https://northfaceenergy.co.za/flight-to-the-future/.

Droughts are not a new experience to these regions. Other regions in Africa that face drought are countries like Sudan, Uganda, Zimbabwe, Angola, Zambia, Mozambique, and parts of Namibia as of 2019. The number of those in need ranges from 2 million all the way to 8 million in these regions individually. Ethiopia is by far the worst hit with some 8.3 million people in need of relief help in water and food supplies. Their state of poverty and horrible living standards leave them to suffer major effects such as malnutrition and famine. Millions continue to die, and the loss of livestock used as their means of livelihood and source of food especially the nomadic communities, leaves them in the dark. Wars and social unrest are also very common as the communities fight over the remaining resources. These conditions leave these regions in a state of food insecurity and high levels of insecurity. In 2011, measles cases broke out in parts of Ethiopia and Kenya (Simelane and Thokozani, 2019).

The Dadaab camp which is a refugee camp confirmed 462 cases and there were some 11 deaths reported. Cases of malaria and cholera also came up reported by the World Health Organization due crowded and unsanitary conditions. The intensity and gravity of the effects in Africa will only continue to worsen if there are no drastic measures that are taken to save the environment. We will continue to count the number of deaths and millions of dollars in relief money to save a condition that can be fixed only by the recovery of the environment and the stopping or reduction significantly of climate change. The domination of poverty, lack of infrastructure, lack of education, complete lack of medical facilities and health facilities are all major situations that only render these areas permanently sentenced to suffer the consequences of major Human activity. The hope for tomorrow and opportunity continues to be a cloud in the sky that slowly evaporates before their very eyes (MASIE, 2020) (Figure 9.15).

Figure 9.15: Tirig, who is six years old, stands with her sister Saua, outside their makeshift shelter in Burao, Somalia. Drought has forced their family to leave home in search of water and food—particularly because the family's goats are now all dead.

Source: https://reliefweb.int/report/kenya/effects-drought-horn-africa-go-far-beyond-hunger.

9.7. LANDSLIDES IN AFRICA

A landslide is a collapse of a mass of the earth oh hillsides or at times rocks from mountains or cliffs. It can at times be referred to as a landslip that is ground movement. They are different depending on the environment of the location that they happen in characterized by steep or slopes from mountain ranges to coastal cliffs or at times even underwater. The driving force to the success of a landslide is gravity. The landslide is mostly triggered by heavy rainfall or an earthquake or even a slope to build a road. These events create great changes of the stability of the ground from stable to unstable in a fraction of time (Figure 9.16).

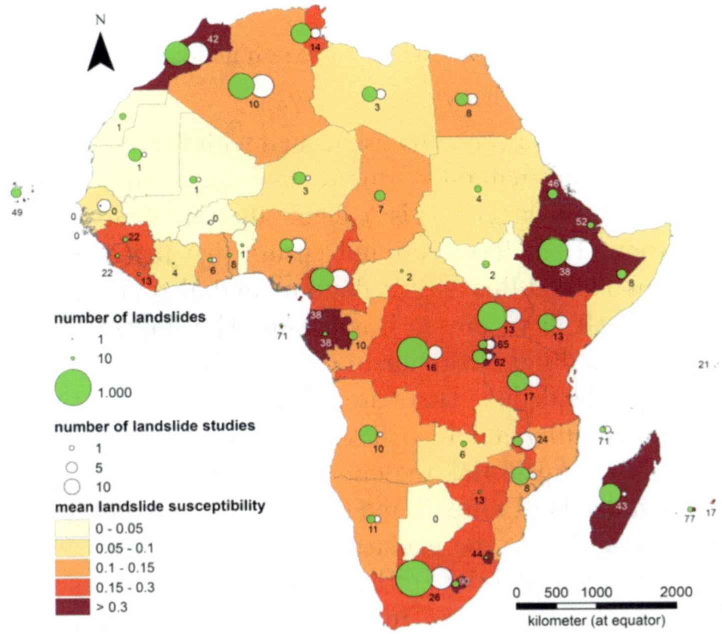

Figure 9.16: A data-based landslide susceptibility map of Africa.

Source: https://www.sciencedirect.com/science/article/abs/pii/S0012825217305494.

They can be caused by naturally occurring events of nature and they can also be caused by human activity. Those caused by natural events include:

- Saturation by rainwater, melting of the snow and or glaciers;
- Increased pore water pressure;
- Erosions;
- Loss of vertical vegetative structures, soil structure and soil nutrients;
- Volcanic eruptions;

- Increased hydrostatic pressures in cracks.

Those caused by human activity are:

- Deforestation;
- Earthwork such as cutting into the earth for road construction or explosive excavation;
- Removal of deep-rooted vegetation that tampers with the structure and stability of the bedrock;
- Urbanization and forest activity such as logging;
- Vibrations from heavy machinery.

Population increase is generally associated with the increase in pressure on the already existing natural resources that are available to man. The regulation of waste management is a factor that is taken very serious in Limbe, Cameroon. They look into and ensure that the drainage systems are well maintained, and this avoids flooding and also is very handy after the events of a flash flood due to heavy rain. The impact that landslides have on humanity and the environment depend almost solely on the speed and the size of the landslide. There are other elements that can come into play such as vulnerability and the risk ratio. In Africa, landslides are mostly experienced in the Equatorial regions. The region is characterized by steep slope, high levels of rainfall, the deep soil profile and the population has a high vulnerability to the geohazards. Annually, there are landslides experienced and they lead to loss of lives, property, as well as affecting the livelihoods of people in a big way. In Uganda, the case studies done there to look deeper into landslides in that region showed that it is characterized by a rural context despite the population density (MASIE, 2020).

The greatest impact faced by Africa caused by landslides if the huge fact that their livelihood, homes, and loved ones is lost in the process. Not being able to make an income in the face of the poverty that they are already dealing with, keeps them barrowing deep in the comfort of poverty. There is also an existing condition of lack of adequate healthcare facilities in these developing countries in Africa. Diseases such as malaria and cholera are still very huge battles that these residents have to face and with the aggravation of the situations by heavy rainfall and landslides, these diseases end up taking root and becoming harder to eradicate (OCHA Services, 2019).

9.8. OVERPOPULATION AND ENVIRONMENTAL DISASTERS IN AFRICAN HISTORY

One of the root causes of an environmental disaster is rapid urbanization. The population numbers in Africa have grown rapidly over the last couple of years. This has brought changes with it of showing a large youth bulge which is reinforced further by the low life expectancy in Africa which is 50 years in some African countries such as Nigeria and Uganda. As of 2017, the population growth rate was 2.6 per annum and the density stood at 44 per sq km. The most populated country in Africa is Nigeria with a growth rate of 2.6% pa at 200 million persons as of 2019. This is followed by Ethiopia with 98,942,102 people, Egypt with 84,705,681 people and Democratic Republic of Congo with 71,264,355 and lastly South Africa with 53,419,333 people. The African population stands at 1,333, 773,819 persons as of April 2020 which is 16.72% of the world's population. The growth rate increases annually by a strong 2%. This population is mostly young people with reports stating that 41% are persons under the age of 15 (OCHA Services, 2019).

There is a population boost in countries that have higher numbers of Asian and European settlers in their countries. Former British colonies such as Tanzania, Kenya, Uganda, and South Africa are the best examples in that scenario. The combination of the high mortality rate and the high fertility rates in Africa contribute greatly to the rapid population growth in the continent. It is said that even if the fertility rate was to decline, there would be no change in the growth rate let alone halt population growth. This is attributed to the "demographic inertia."

The main reason that contributes to the high fertility rate is mainly the cultural practices of not believing in modern contraception or believing in the ideology as a whole. Women have little to no control over their lives and those that would be willing to try birth control, have zero to no access to it (United Nations, 2010).

Less developed countries in Africa depend mostly on farming as a way to make a living. Children are often exposed to work at an early age exacerbating the child labor rates. The notion that drives this is that the more children, the more the workers, the more income one gets. The traditions such as wife inheritance and polygamous marriages contribute largely in the making of large families that are more than likely not sustainable. Sex among couples is practiced more often and with no birth control, childbirth is inevitable. Abortions are not readily available in the poorly managed healthcare facilities that are available. The concept of abortion is also frowned apron by the communities hence leaving women with no choice but to bare more children. There is also a notion of the male child as a validation for male virility. Young girls are married off at young ages and they start having children as early as the age of 14. In the event the girl has a girl, they will try and try until there is a male child in the family. This is practiced in many of the countries in Africa and leads to larger families and many families numerically, in the long run contributing to the population growth.

The high population numbers with the youth being the majority, the rate of immigration to other countries and continents in search of greener pastures is very high. This is caused mainly by the lack of employment in their home countries. Some do not have the education to pursue their career dreams and are left to do odd jobs in developed countries fir income. They send money home as to help those they left behind still struggling with poverty. Those that have the education lack jobs due to nepotism and corruption which is likened to a wound that will not heal in Africa. The national economies continue to stagnate, and they are left to outsource for workers and developers for their own countries which costs billions of dollars.

With the population only growing, the strain on environmental resources continues to grow as well. The communities continue to struggle in dividing the available resources Among the dense population. This creates room for risk factors for disasters and the continent remains vulnerable. Drought, landslides, famine, erosion, flooding, plagues, sandstorms, deforestation, mining, and more environmental disasters are only waiting to happen due to the great number of human activity and pressure exerted to the environment to boost production for to meet demand. There will be continuous land degradation, and this will create bigger holes in the environment leading to climate change (Babar, 2007).

9.9. HEAVY RAINFALL AND FLOODING IN AFRICA

Flooding and heavy rainfall is more dominant in the East African region. Kenya, Ethiopia, Uganda, Rwanda, Somalia, Djibouti, and Rwanda experienced one of the deadliest floods in 2018 affecting millions. This was preceded by severe drought that was met with heavy rainfall, massive landslides and flooding as well as saw the overflowing of several dams. The impact countries suffer from severe loss in property damage and loss of lives by their thousands in every disaster happening. In Rwanda, flooding that happened in Rwanda hills brought about several landslides that killed about 200 individuals. In May of 2018, the Kigali-Gatuna highway which is one of the major roads into and out of the city Kigali was washed away by flooding and this disconnected many vital trade routes for the landlocked country. The damages incurred by the destruction of this road tallied up to $28 million (Dufresne, 2019).

In Kenya, at least 186 persons died in 2018 due to landslides, buildings collapsing, capsizing, and drowning all brought about by heavy rainfall and flooding. The displacement of some 280,000 people and families takes a toll on the struggling nation. The 2018 floods contributed greatly to the failure of Patel Dam leading to the 48 fatalities. There were four additional dam failures and bursts in Lake Baringo region on May 26th, 2018, claiming the lives of 5 people and displacing another 10,000 individuals. Damages in repair for the country were a total of $187 million (Figure 9.17).

Figure 9.17: People stand on debris blocking a highway after River Muruny burst its bank following heavy rains in Parua village, in Kenya's West Pokot county.

Source: https://www.voanews.com/africa/east-africa-struggles-heavy-rains-more-come.

Flooding may come as a result of heavy rainfall and that overflows from the designated channels and at times does not even reach the defined channels. The heavy rainfall causes ground water levels to rise to the surface thus releasing a lot of water that floods. There is also a huge contribution from poor river basin management in the areas prone to heavy rainfall and eventual flooding. There is also no local management of drains and sewers that help with the draining of rainwater and directing it to designated channels. This causes the water to overflow and cover areas of land that cannot help control water movement. This increases the risk in cities and towns. With human activity in more open areas with not too much infrastructure, deforestation, excavation with the use of explosives, erosion, we weaken the soil and the ground is more than willing to give way when there is heavy rainfall for flooding to occur (Babar, 2007).

The damages financially and economically for those affected can only be too great. There is an increase in the number of fatalities and the death toll keeps shooting up. The loss of livelihood in terms of crop produce, animals, and buildings of work demoralizes the spirits of those affected. The open opportunity for water-borne diseases such as diarrhea, malaria, typhoid, and cholera take the root of the day due to water contamination, poor hygiene, poor sanitation, and creation for breeding grounds for pests and parasites. The birth of new diseases is also a huge risk factor especially in areas like these that have poor health facilities and the likelihood of poor immune systems by the people of the community due to poor feeding and malnutrition (United Nations, 2010) (Figure 9.18).

Figure 9.18: Rainfall from October to mid-November has been up to 300% above average in the greater Horn of Africa region.

Source: https://www.aljazeera.com/news/2019/11/east-africa-struggles-heavy-rains-thousands-displaced-191129061914721.html.

There is also an influx in various sectors of the economy. The cost of living goes u. production of food and the retail price of food also shoots up significantly. The transport and tourism industries also suffer a great deal due to the drop of number in tourists and transport passengers. Internal community wars and violence comes up as they fight and scramble for the only available resources for their families bearing in mind the heavily populated numbers in Africa.

9.10. SWARMS AND DEADLY DESERT LOCUSTS

Locusts are collection upsetting species of the grasshopper family. These insects are usually solitary but due to different circumstances and situations they can become Large in number and change their behavior and habits. These grasshoppers are normally low in number and pose no threat to the economy or the agricultural well-being. In situations of drought followed by high vegetation growth, they are triggered and cause a dramatic change in their breeding behavior. Locusts have formed plagues since prehistory. There are records from the Bible and the gram about the ancient Egyptian swarm of locusts. The storms have devastated crop production and are catastrophic causing famine and human migration. The changes in agriculture in recent times a better survivor lands of locations by which the swans tend to orient and also have control of the swarm signs in early stages (Simelane and Thokozani, 2019).

The swarming behavior discussed in the 20th century had decreased but over the years despite modern civilization and development, Potential swarms can still form depending on location and the suitability of the climatic conditions. Locusts are large insects and they are used in research and in the study of zoology the classrooms (Figure 9.19).

Figure 9.19: A local guide holds a handful of dead locusts at Shaba National Reserve, Kenya.

Source: https://www.aljazeera.com/indepth/inpictures/pictures-locust-outbreak-spreads-east-africa-200120073558008.html.

Locusts can be said to be the swarming phases of certain the short-horned grasshopper in the Arachnidae family. They are often solitary, and the no taxonomic distinction made between the locusts and the grasshopper is the fact that the swarming kind forms these large numbers under intermittently suitable conditions. In English terms, locusts are the grasshopper species that have a change in behavior under the right climatic conditions and they change morphology and behavior and they start to crowd and form swarms that develop from brands of immature stages of the hopper's lifespan. These are the world's most dangerous migratory pests. They can reach numbers of up to a billion in a single swarm which poses a serious threat to the vegetation and crop produce for man.

Over the last Millennia, locust plagues have been a part or man's history. In the 1875, Albert's swarm was estimated to have covered 198,000 square miles which is 510,000 km squared. The weight was estimated to be about 27.5 million tons with 12.5 trillion insects (Simelane and Thokozani, 2019) (Figure 9.20).

Figure 9.20: Katitika village, Kitui county, Kenya-A swarm of desert locusts feeding on crops.

Source: https://www.unenvironment.org/news-and-stories/story/locust-swarms-and-climate-change.

The African migratory locust is the most common locust in Africa with the behavior of swarming. In the African south of the Sahara Desert, the breeding conditions there are favorable making the Sahara Desert the main breeding ground or these locusts. The Niger River in West Africa is the main source of the plague. Many plagues of this kind took place between the years of 1891 to 1903 and 1928 to 1941. With recent climate change, there have been more than usual outbreaks of these plagues all over Africa. As of 2020, the affected countries are Somalia, Kenya, Ethiopia, Eritrea, Uganda, and Tanzania. In Northern Kenya, the swarms are large in number and measure about 35 miles long and 25 miles wide radius. They contain up to 80 million locusts and they attack land with vegetation. In the recent times, the locusts have flown into South Sudan and they have accepted the levels of hunger. Famine and starvation are something that is a serious result of the locust swarms due to the consumption rate of these locusts on the crop lands they infest on. They consume roughly about 2 grams each per day in the desert. If we have a swarm of millions or trillions, there can be consumption of tones of crop and vegetation on the land leading to no food or vegetation for the environment to function with or for humanity to feed on.

The impact faced by the affected individuals is lethal to the population and the sustainability of life. With the locusts consuming a minimum or what is equivalent to 25,000 people per day in each swarm and more, the food security remains in harm's way. The environment also suffers great damage as vegetation is key for the eco-system to function as normal. There is little to no crop production for earning a living for African families that depend of farm produce to earn a living. With the population starving and the productivity levels going down, the economic income and stability in these local communities affected goes below average and there remains zero room for development and recovery from the current citations (IFRC, 2011).

National governments incur a lot of costs in the buying of the treating chemicals used to spray the swarms in motion and in more than often scenarios, the locust's resilience pays off and they remain alive and continue to consume. That renders the spraying chemicals weak and ineffective. The government is left with no choice but to procure more chemicals in the hope that they will work. These chemicals have an effect on the crops in the farms and they at times cause poisoning of the crop produce causing great havoc and likely death of its citizens that consume the crop produce (IFRC, 2011) (Figure 9.21).

Figure 9.21: Desert locusts cover branches in Katitika village, Kitui county, in Kenya on Friday, January 24. Kenya hasn't seen locust swarms of this size in 70 years.

Source: https://www.smithsonianmag.com/smart-news/billions-locusts-are-swarming-east-africa-180974135/.

The measure countries can take to be better prepared for these similar events is to stop climate change. The conditions for the breeding of these locusts are created by the effects that come about the climate change results all over the world. The development of climate change solutions should be an implementation that local governments in Africa should take seriously and educate the people on the best ways to make the environment better and preventive measure for climate change. The United Nations has a response team that deals with the prevention and the control of these pest invasions.

Of all the topics covered by this chapter, one message stands out, climate change. This concept is an important factor that determines the health of the environment that we have. The normal functioning of the environment was taken away years ago by the urbanization and development introduction into civilization. The greed of man has since been the fuel for the deteriorating state and condition of the environment. Deforestation in Africa, gully erosion crises in Africa, earthquakes in Africa, cyclones in Africa, droughts in Africa, landslides in Africa, pollution disasters in Africa, heavy rain and flooding in Africa, and swarms of deadly desert locusts are only but a few of the environmental disaster that the continent can deal with. If the conditions of the environment do not change, there could be experiences of tsunami like disasters and worse faced by the continent.

The fact that there is not enough funding for the emergency response teams that handle these types of disasters, tells a lot about the economic preparedness in the continent. There are also very few functioning and well-equipped healthcare facilities that offer medical help to the citizens. The rise of population speaks volumes about the need for civic education for the general public that is considered vulnerable and at great risk of these disasters. The education on basic hygiene, sanitation, birth control options, management of waste and water sanitation is key to boosting the living conditions and standards of these people.

The development of the environmentalist ideology also creates hope for the health of the environment. This would bring forth the need to plant more trees, completely stop deforestation and logging, prevention of pollution of water and air, industrial regulations on the wastes they produce, the ban on plastic manufacturing and usage, fines, and strict measures for those participating in bumping, preventive measures for soil erosion among many other measures that could be generated locally to create a social movement to rehabilitate and restore the environment. The merging of local organizations and Non-governmental organizations with world bodies that are involved with environmental conservation and restoration to be part of the movement and involving the local communities can be involved and participate on a first hand basis of the ways to save the environment and the need to save the environment will be hammered deeper into their minds. This will bring up the spirit of the individuals participating and slowly build back the agricultural structure and eventually the economy in African localities will boost up setting them up for development and recovery from poverty, this being the main chain holding them down, years after civilization, years after colonization and the concept of urbanization being introduced. There will also be a creation of livelihood for the communities that create more jobs and possibility for a better future for them.

9.11. REAL LIFE EXAMPLES OF ENVIRONMENTAL DISASTERS

9.11.1. The Oil Spills of the Niger Delta of Nigeria

Niger Delta is a vast region consisting of rich mangroves as well as fish-rich channels. Many residents seek to make their daily livelihoods from farming and fishing activities. However, huge oil spills are quite common in Niger Delta where more than 40 million liters of crude oil get spilled annually, causing human deaths, and destruction to the local environment. This threatens the livelihoods of those who rely on the Niger Delta for subsistence. A 2018 inquiry by the Health and Pollution Journal discovered that well over 12,000 oil spill cases have happened in the oil-rich area between 1976 and 2014.

9.11.2. Merriespruit Mine Slick Disaster, South Africa

On February 1994, Merriespruit tailings dam in South Africa failed after overtopping as a result of heavy rains triggering a flow drop (static liquefaction) of section of the embankment. By large, water mismanagement was blamable as it caused around 600,000 m^3 of tailings or (1.2 million tonnes) to be mobilized from the impoundment where water flow eventually ceased 2 km away in Merriespruit town.17 people perished and scores of homes demolished in the aftermath.

9.11.3. Ivory Coast Toxic Waste Dump, 2006

Roughly, 15 years ago, a cargo ship known as Probo Koala reached termination of a 4-month journey which resulted in hazardous waste being discarded illegally in Ivory Coast. The huge scale of the effects of the dumping caused headlines throughout the world. Well over 100,000 people required medical aid. The culprit was a multinational oil trading firm Trafigura, which produced the lethal waste aboard the ship due to refining of a tainted petroleum product known as Petroleum naphtha.

9.11.4. Deforestation in Nigeria

It's believed that Nigeria loses approximately 1.5 million trees per day due to unlawful logging and other contributors. Both human and natural elements such as global warming, desert encroachment, insurgency, over grazing and so on are to blame. Deforestation has many negative consequences on the environment with regards to soil erosion, loss of wildlife, heightened desertification, and loss of biodiversity systems.

9.11.5. Draining of the Aral Sea

Aral Sea is located in Central Asia between the Southside of Kazakhstan and North Uzbekistan. Up to third quarter of 20th century, it ranked as the world's 4th largest saline lake, containing around 10 gms of salt for every liter. The Soviet government however decided during the 1960s to divert rivers feeding the Aral Sea, so that they may irrigate the desert area around the Sea for agriculture. This caused drying of the sea which affected the health and lives of several millions of inhabitants around the region.

10

Worst Man-Made Environmental Disasters in American History

CONTENTS

10.1. INTRODUCTION

A natural disaster occurs by a natural calamity that results in tremendous damage or loss of life. We mainly link the term to tsunamis, earthquakes, and typhoons. As much as it comes very easy to attribute the meaning and cause of the disasters to natural phenomena on Earth, we as human beings are responsible for some of the most horrendous and dangerous act upon ourselves and nature either deliberately or by accident. Man-made disasters is a part of the human process and experience and although many have happened that resulted in no deaths or injuries there is a considerable number that have. In relation to this writing, we will focus on disasters that have been attributed as a result of man and their activities that damage the environment. It is also possible for a natural occurrence and disaster to be turned into a man-made environmental disaster and that will be covered as well.

In most part, environmental disasters are actually caused by human error either by accidents lack of foresight corner-cutting and greed in the commercial world or basic ineptitude and incompetence. Natural disasters in themselves are natural and therefore in the destruction there is a developmental process that the environment must and will take. However, with man-made environmental disasters, the disaster itself will attribute to harm and damage in the environment and that in turn increases the risk to human beings and the process can become cyclical. Climate change, whether theorists may attribute it to man-made or natural causes is a catalyst in and of it and will increase and intensify the relationship between disasters the environment and man. In any case, however, nobody can deny that the impact of human altercations on the ecosystem has being long lasting. These impacts that we have had are in relation to various environmental processes and systems including topological geomorphological atmospheric geochemical and hydrological among others.

The United States of America has a very rich history and plenty of it has been a story of overcoming catastrophes in as much that many of them are self-induced. Just as in any other places, you will find that man-made environmental disasters in the US will be most affective to the poorest in the population. The nature of human development means that they are most vulnerable to environmental degradation settlement patterns livelihood choices and behavior in general towards the disasters risk. On top of that, the poorest also are the least resilient in terms of meeting recovery needs from environmental disasters because of their overall lack of support network, insurance and alternative livelihood options. This chapter aims to give a comprehensive approach towards man-made environmental disasters more especially the worst of them in the history of the United States of America and with a specific look towards these disasters and how they affected the ecological systems and the environment. This is because we must recognize that the environment itself is vulnerable two disasters of all kinds and even more especially man-made disasters.

Because of the fact that acts of war and terrorism are not an accident but intentional and deliberately driven towards affecting other human beings directly; and that they are man-made purposefully orchestrated disasters to single out man, they are not quantifiable in this assessment. September 9/11 for example was a terrorist attack that was in all truths a disaster but it had minimal effect on the natural environment and ecosystem. It took away the lives of 2,996 people and that cannot be understated however, it was not an environmental disaster and had minimal ecological effects in as much that it was a very man-made orchestrated calamity. This is the primary reason why the nuclear weapons detonation at the Nevada test site will not be assessed in this writing although it had major environmental consequences.

Ranking disasters is usually a factor of monitoring the death toll first as human lives are immeasurable in value (automatically this means that the disaster that kills the most people is the worst). Aside from that, the amount of damage estimated in monetary value is the next step in grading the severity of an environmental disaster. The third criterion used for the purposes of this chapter was the level of Environmental and ecological damage from wildlife killed and animal habitats destroyed. Without much more debate, let us take a look at the worst man-made environmental disasters in American history.

10.2. THE PESHTIGO FIRE OF 1871

The Peshtigo fire is the most devastating and destructive fires of American history that swept through 1.2 million acres or 490,000 acres from northeastern Wisconsin around the Green Bay area covering much of the Southern Door

Peninsula and adjacent areas of Michigan. Peshtigo Wisconsin was the largest community that was affected by the fire and it is estimated that the total number of fatalities was anywhere between 1500 to as many as 2500 people. The reason there is such a huge variance in the estimated death toll is due to the fact that many of the records were actually burnt in the fire.

The financial damage was estimated to be about 170 million dollars. The fire burnt nearly 20 other towns but by far pushed eagle was the worst hit with 800 lives lost and the City evaporated within an hour. It had such a damaging and horrific effect on people's lives and psyche that many thought it was the end of the world.

There are several theories for the cause of the fire but the most convincing so far was the preceding hot summer paired with farmers and agriculturalists setting controlled fires to make room for the new planting season. Together with farmers trying to clear Forest land and also the setting of small fires to make room for railroad construction, either or both could have been the reason but generally it is agreed upon that it was an environmental disaster of human making. The small land-clearing fires or at least one of them must have blown out of control and created what is known as a firestorm.

A firestorm is a conflagration which builds up and grows wildly to such intensity that it creates and maintains its own wind flow systems. In fact, it is a phenomenal that is characterized by strong gale like storm force winds that blow towards the fire from everywhere around its perimeter creating a column of hot gas over it that draws in cool air. This allows anything in the periphery to be dried up that allows even more intense ignition. The interesting thing is that with firestorms unlike large wildfire conflagrations we will see strong pressure gradients that are more or less formed by a number of the ignition sources rather than just one. Intense flame fronts can be built especially if there is some sort of emission of combustible gases like methane explosions which was exactly the case in the Peshtigo fire. The fire was so intense that its convulsive speeds consumed available oxygen with some trying to flee bursting into flames. The only way the fire goes out is by running out of fuel which in any case everything burns in temperatures excess of 2000°C in hurricane force winds (Figure 10.1).

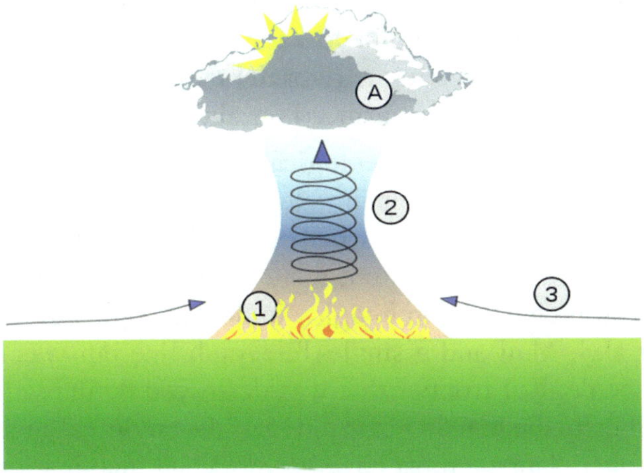

Figure 10.1: Above is a schematic view of a firestorms thermal column. Fire (1) makes hot air (2) to rise due to pressure forming storm-winds (3) towards the fires. Above it, a pyro-cumulonimbus (A) can be observed.

Source: https://en.wikipedia.org/wiki/File:Firestorm_thermal_column.svg.

The most intriguing fact about this environmental disaster is that it happened on the same day as the great Chicago Fire that was a much more famous urban conflagration the killed approximately 300 people and destroyed roughly 9 square kilometers of the city. On top of that was the Great Michigan fire as well that was a series of simultaneous forests conflagrations. It is possible that wind patterns and systems made all these separate wildfires propagate so widely but it can be agreed that Peshtigo was the deadliest wildfire in American History despite its lack of infamy. Port Huron at the southern end of Lake Huron also had major fires on the same day which leads to many contemporary conspiracy theories and historians alike to assume they had a common cause; like the speculation of the Comet Theory. The comet theory stipulates that Fragments of Biella's Comet scattered all over the northern end of the United States igniting the several but somehow linked infernos however this was quickly debunked because meteorites cannot ignite a fire because the fragments are too cold when they touch the earth surface.

The Peshtigo fire museum is located in an old former church building in the city where various Relics and records about the disaster remain, for example, a story of a man carrying a woman to safety that he thought was his wife that went mad when he found out later that she wasn't. Another interesting story involves a young German girl of 13 years who floated on the Peshtigo River on the horn of a cow for several days in order to survive the fire. It recorded that once in the river the faces of those trying to escape where scorched each time they surfaced for air.

From an environmental and ecological perspective, an entire ecological system is damaged beyond repair. New Forest management programs and lumber harvesting techniques had to be developed to ensure further damage to the environment was not done because of forest was no longer an unlimited resource. One can imagine in the late 1800s the area was only reachable By U.S. Mail and a single Telegraph line but was otherwise an isolated wild ecological frontier that was destroyed by the wildfires. Many records only dictate the human perspective of the environmental disaster but the true ecological damage must have been beyond imagination.

10.3. THE DUST BOWL OF THE 1930S

Dust bowl of the 1930s is a great example of what the consequences of over-aggressive farming can do to the environment. History begins with the Foundation of the United States where the Federal Government encouraged settlement into the semi-arid area offering 160 Acre plots to settlers (through the Homestead Act of 1862). The area receives less than 50 centimeters of rain annually and is a largely short grass Prairie biome spending to the Rocky Mountains extending to the Rocky Mountains and is fairly unsuitable for European Style farming most people moving to the area (largely explorers) calling it the Great American Desert.

The United States government expanded on the 160 Acres offering 640 Acres instead to potential settlers. Unusually wet weather in the beginning of the 20th century encourage waves of European settlers who's hotter the semi-arid area could now support large-scale agriculture. Technological advances that allowed large-scale farming like mechanized ploughs and harvesting methods made the offer more attractive especially for large operations without increasing labor costs.

The Russian Revolution and World War I decreased the global supply of crops like wheat and other commodities while increasing agricultural prices respectively. This exponential demand for food encouraged farmers like those on Llano Estacado to expand their farmlands and dramatically increase their cultivation. The native graph the held the soil together and then please as well as allowing moisture attention was eliminated on a large scale. Deep plowing also promoted large scale soil erosion and during winter months, the fields were left bare when winds around the area were highest.

This insufficient understanding of the Plains ecology and the Farmers over aggressive farming created the worst conditions and were exacerbated by the droughts of 1934 1936 and 1939. Some regions experienced drought conditions for as long as 8 years. The soil had nothing to Anchor to and turned into dust that blew with the prevailing winds sometimes turning days into pitch black. These choking dust storms name the black blizzard would travel across the country reaching as far as the East Coast and reducing visibility on the planes to less than 1 meter. The term the term dust bowl was coined by an Associated Press reporter who happens to be in Oklahoma at the time and experienced a black blizzard on April 14th 1935. The dirty thirties as it is sometimes called affected a region of 400,000 km squared (100 million acres of land) and forced tens of thousands to abandon their farms. Disaster led to widespread hunger and poverty with more than 500,000 Americans left homeless. This led to the largest migration in American history with 3.5 million people having moved out of the plains. 6500 people were killed during only one year of the dust bowl some while trying to hop on freight trains to get to other parts of the country

In 1932, there were 14 dust storms and in 1935, there were 38. Many people many people driving away how to drag chains behind their cars because the dust bowl caused such high static electricity that it was the only way to keep them grounded otherwise they would suffer severe electric shocks from whoever from touched other people or objects (Figure 10.2).

Figure 10.2: A dust storm approaching during the dirty thirties.

Source: https://vegetablegrowersnews.com/article/jerry-mills-remembering-the-dust-bowl-climate-of-the-1930s/.

Government responses well paid with many different groups to identify different approaches to managing the disaster and groups such as the soil conservation service generated details soil maps and took photos of the land from up above. Others like the United States forestry service created shelter beds to reduce soil erosion and Franklin D Roosevelt first 100 days in office quickly initiated programs to conserve soil and restore ecological balance not only to the area but the entire nation you recognize this disaster as one of the primary causes of the Great Depression. The economic the economic effects persisted because some farmers also failed to switch to more appropriate crops and made limited adjustments to their farmlands. It made such a great impact on the American people that it had an influence on the arts and Culture with examples like the 2014 science fiction film interstellar having comparative features and American recording artist Grant Malloy Smith releasing an album known as Dust Storm-American Stories in 2017.

10.4. PICHER OKLAHOMA LEAD CONTAMINATION (1917)

Since the early 1900's Oklahoma has been one of the biggest mining towns in the United States. Between 1917 and 1947, just 30 years 20 billion dollars' worth of lead and zinc has been mined there. It was a thriving town with thousands of people working in the mines and support services however, toxic waste was piling up in the area, and the water was beginning to carry a reddish-brown hue. More than a century of excavation left giant piles of toxic metal contaminated chat throughout the area. A 1996 study found that 34% of the children in the area suffered from lead poisoning due to leaks of lead and zinc that could result in lifelong neurological conditions. A 2006 study showed that 86% of the buildings War in such bad shape and likely to collapse at any time and the e p a together with the state of Oklahoma agreed that the entire town should be evacuated mandatory. The town's municipality was officially dissolved in 2013 (Figure 10.3).

Figure 10.3: Lead contaminated water in Picher Oklahoma.

Source: https://owlcation.com/stem/-Ten-Worst-Man-Made-Environmental-Disasters-in-American-History.

Today Picher is a ghost town in every sense of the word and is totally unlivable it is actually safe to visit but you should stay on paved roads and the area is susceptible to cave-ins. Together with the Love Canal and Kotzebue Alaska in the Arctic Circle, Picher Town is deemed one of the most toxic areas in the US. Surprisingly January 2011 counted six homes and one business remaining with their owners refusing to leave even at any cost. The city residents that moved away come back to the town for Christmas parades which begin from the high school and down the main street. The last inhabitant of Picher was a pharmacist known as Gary Linder man who vowed to stay as long as there were people to serve there. He died in 2015 officially dropping the population of the town to zero.

Man-made environmental disasters as one can analyze from the above case studies are primarily due to negligence and incompetence and very preventable. Many disasters even though they may appear natural at first have an underlying human Element that furthers and intensifies the negative effects that they carry Disasters however, natural or man-made, should remind us and reiterated the importance of preparation and being constantly informed about our ecological systems and human development on both a very local and bigger national scale. Our inseparable relation with our environment should be counted more of as a blessing rather than a challenge to be overcome with efforts of conservation and human development having a bigger picture on how we conduct ourselves. It is unfortunate that we only have a brief insight on the importance of stewardship of the environment only when a disaster hits and shortly afterwards the earlier greed and power short-sightedness returns. A mental and behavioral change is required not only to minimize the effects but altogether prevent man made environmental disasters.

10.5. TEXAS CITY DISASTER (1947)

The Texas City disaster was one of history's largest non-nuclear explosions that happened in the Port of Texas city at Galveston Bay. Ammonium nitrate was manufactured and transported by rail from Nebraska and Iowa to Texas City. The 2100 metric tons the ship was carrying was needed either as fertilizer or explosives and was to be loaded onto SS Grand Camp, a French registered vessel.

The ship was originally named SS Benjamin R Curtis and served in the Pacific Theatre in World War II as a reserve Fleet cargo carrier. The 133-meter-long ship was assigned to the French line as a cold war gesture in assisting to rebuild France along with other efforts. Together with the large load of ammonium nitrate, the ship was also carrying ammunition for small arms, machinery, and bales upon bales of sisal twine. At the time of the disaster another ship in the harbor SS High Flyer was docked not more than 200 m away, carrying 872 metric tons of ammonium nitrate 1600 tons of sulfur

The ammonium nitrate was packed with clay petroleum and paraffin wax and layered in paper stacks. Before loading the dock workers and longshoreman reported that the bags felt warm to the touch and by 8 a.m. smoke was spotted on the cargo hold. Over the next hour, the crew tried to extinguish the fire but failed as a red glow Returned each time they made an attempt. After that the captain ordered the men to use a fire fighting method where's team is piped into the holes to extinguish the fires to preserve the remaining cargo. This attempt also failed because ammonium nitrate produces own oxygen and continues to burn neutralizing the extinguishing properties of the steam.

Meanwhile the dock was attracting spectators who assumed they were at a safe distance. Little did they know that steam pressure was building inside of the ship which eventually blew open hatches and yellow orange smoke plumed out, typical of any nitrogen oxide fire. The bright colors of the smoke attracted even more spectators who noticed that the waters around the Dock of beginning to boil from the heat. The entire cargo holds of the ship and the deck as well began to bulge and buckle under the pressure as the steam inside increased. At 9:12, the explosive threshold was reached due to the combination of heat and pressure and the vessel detonated.

The blast was so destructive that shock waves 5 meters' high were detected almost 160 km away from the Texan shoreline. It destroyed nearly 1000 buildings one of them being them and the Monsanto Chemical Company plant that produced and housed agrochemicals and biotechnology. Hundreds of falling bales of twine contributed to the damage and the two-ton SS Grand Camp anchor was thrown more than two and a half kilometers across the city together with almost 5800 metric tons of the ship's steel, some at speeds beyond supersonic. Planes that were flying nearby had their wings shorn off and fell from the sky. Half the windows in Galveston town which is over 16 km away were shattered. Seismographs measured the blast as if it were an earthquake (Figure 10.4).

Figure 10.4: SS Grandcamp 2-ton anchor was thrown more than 1.6 miles when the exploded, now stored in a museum.

Source: https://en.wikipedia.org/wiki/Texas_City_disaster#Explosions.

An estimated 567 people lost their lives that day with more than 5000 people injured and 2000 left homeless. The whole 28-man Texas City fire department was deployed to fight the shipboard fire and the initial explosion killed all of them but one. The explosion ignited the SS High Flyer carrying even more aforementioned amounts of ammonium nitrate. The crew spent hours trying to cut the SS High Flyer from its anchor to never her away but failed and more than 15 hours after the SS Grand Camp was destroyed the SS High Flyer exploded killing more and damaging the port and even other ships like the SS Wilson B Keene (Figure 10.5).

Figure 10.5: A styrene plant became a roaring inferno on April 16, 1947, in Texas City. Hundreds of other fires started in areas where petroleum was stored.

Source: https://www.houstonchronicle.com/neighborhood/bayarea/news/article/The-1947-Texas-City-Disaster-devastated-the-Bay-11071630.php#photo-10411673.

The most immediate environmental effect was the burning it the adjacent areas and an oily fog that left deposits on every inch of every outdoor surface in areas as far as Galveston. Oil storage tanks and other chemical storage facilities including a five story rubber factory and a styrene plant (where polystyrene foam and plastic are made) were ignited as well and the man-made chemical conflagration polluted the air. Property damaged was 1.1 billion dollars accounting for inflation. The disaster resulted in the first ever class action lawsuit against the US Government for the benefit of the 8485 victims on the grounds on litany of negligence and 17 million in damages was paid out. This man made environmental disaster is ranked the as the worst industrial accident in the history of the United States.

10.6. THE LOVE CANAL TRAGEDY (1978)

The love canal tragedy was one of the most horrendous environmental tragedies in American history. Like some of the above stated environmental disasters, it was not exactly an isolated event. The story begins with William T Love who had a dream of building a community around a three block tract of land on the Eastern edge of the Niagara Falls in New York. William Love he was an ambitious entrepreneur from the Western railroad corporation and organized the construction of a short canal between the upper and lower Niagara Falls for the generation of cheap power that would be supplied to his small dream town. Mind you all this was happening in the beginning of the 20th century when Nikola Tesla's discovery of how to transmit electricity on an economically viable way across great distances was only beginning to yield fruit. Despite considerable support, the fluctuation of the economy (the panic of 1907) and various other setbacks Loves Project was never able to become a reality. By 1910, the project was all but abandoned and all that was left to commemorate the Model City dreams was a partial ditch where the construction of the canal had started. The canal filled up with water and the local children would skate over the frozen waters during the winters and swim during the summers. This may be a life threating disaster, especially if nothing is done to mitigate the disaster .

In 10 years' time, the canal now had become a municipal and industrial dumpsite that was totally unregulated. Landfills are actually an environmentally acceptable method to dispose of all sorts of waste assuming the area is continuously and deliberately inspected and managed well. In the 1940s, the Hooker Chemical Company was desperately looking for a place to dispose of large quantities of industrial chemical waste. Initially when the company was formed they manufactured and produced sulfur chloride and sodium chlorate and where primarily into the hydrogenation of vegetable oils. By 1940, the American chemical company was a big producer of chlorine that was used in the chlorination of drinking water sold as chlorinated lime. They also produced chlorobenzene that, through a chemical process can be converted to picric acid that is used in the making of explosives. They also produced phenol and trichloroethylene that is sold for degreasing and the used laundry and dry cleaning industry.

Meanwhile the population around the Niagara Falls was growing rapidly. Schools and entire Communities were being constructed around the area and by 1952 when the Hooker Chemical Company realized that the site would likely be developed for further construction, they stopped using it as a dump site. This is not before they had buried around 20,000 tons of caustics, alkalines, fatty acids, and chlorinated hydrocarbons at a depth of 6 meters on average. The area was sealed with clay to prevent leakage.

The Niagara Falls school board approach the Hooker Company to purchase The Love canal property for the construction of a new school and. Rocket Chemical Company considered the proposition and eventually sold the property to the city for $1 although the school condemned some nearby properties. They felt that it would be a wise decision to turn over the property to the schools provided they would not be held responsible for future claims and liabilities resulting from the Underground storage of chemicals. This was a bad transaction.

The school was built and about 100 more homes right on top of the industrial chemical disposal site and for a while, it functioned as a solid working-class community. Little did they know they were living on top of a ticking time bomb. In 1978, the area experienced a record amount of rainfall and very wet winters and shortly thereafter leaching began. Residents began noticing suspicious black fluid flowing out of the love canal and strange odors and substances in the yards in public playgrounds. A series of inexplicable illnesses; epilepsy, migraines, and nephrosis together with an abnormally high rate of birth defects and miscarriages swept the love canal neighborhood by storm. What had happened is the water table was raised and underground swales and sewer systems that drain into nearby creeks were contaminated (Figure 10.6).

Figure 10.6: Love canal residential areas being excavated for clean-up.

Source: https://www.earthshare.org/lovecanal/.

All these concerns were very alarming and the City eventually hired a consultant, Calspan Corporation a science and technology company a science and technology company and part of the research laboratory of the Curtis-Wright airplane division in Buffalo New York. They discovered up to 82 different compounds, 11 of the suspected carcinogens, that had been percolating up into the school basement of homes and playgrounds where young children would return home with inexplicable burns on their hands and faces with the ever-present choking smell in the air.

The human impact was tremendous. For a tiny community to report five birth defect cases was terrifying together with a high rate of miscarriages. Large percentage of people will also closely observed due to the High white blood cell counts which is a possible precursor to leukemia and other forms of cancer. President Jimmy Carter declared a state of emergency and ordered the Federal Government to relocate over 200 families. This was not enough as over 700 families warranted live location due to the toxic substances that were leaching into their homes. A second state of emergency was declared in 1981 where the remaining families were relocated and the total cost for cleanup was 257 million dollars. The Hooker Chemical Company changed its name to the Occidental Chemical Company and in 1995 paid 102 million dollars to the Superfund that was created for the cleanup. The Superfund as it is better known was formed in 1980 as The Comprehensive Environmental Response, Compensation, and Liability Act went on to analyze tens of thousands of hazardous waste sites in the US and assisted in the cleanup of hundreds of the worst ones (Figure 10.7).

Figure 10.7: Protests of residents at love canal gained global media attention and fueled many environmental justice movements.

Source: https://www.pinterest.com/pin/527765650075076913/?d=t&mt=login.

The love canal stands as a significant Turning Point the history of environmental justice and over the past several decades the number of environmental justice movements (especially grassroots militant ones have grown globally with numerous advances in legislation and activism. Terms like environmental racism where marginalized and poor populations would house chemical factories and industrial toxic chemicals and waste were brought to use by researchers and activists because of the Love Canal Incident. The disaster symbolized the impending environmental calamity that is represented by hundreds of toxic waste disposal sites spread across the United States and also brought into light the failures of mainstream environmentalism and mounting frustrations of the public.

10.7. THE EXXON VALDEZ OIL SPILL (1989)

Until the Deepwater Horizon oil spill of 2010 the Exxon Valdez oil spill was the worst oil spill in US history. The Exxon Valdez was an oil tanker owned by the Exxon shipping company that was headed to Long Beach California from Prudhoe Bay oil field that is located just off the Alaskan North slope and his largest oil field in North America. The ship struck Prince William's Sound Bligh Reef that is a well-established navigation Hazard just a little past midnight. The massive 300 meter long, 51 meter wide singe hull tanker ripped open its belly on March 24th 1989, pouring 11 million U.S. gallons of its total 53 million gallons of crude oil into the open sea. The Captain Joseph Hazelwood had been drinking at the time and allowed the unlicensed third mate to steer the massive ship.

Interestingly enough although it was not the largest it is considered by many experts as the world's worst oil spill in terms of damage to the environment and marine ecosystem due to the difficulty to access the area where the accident occurred. This situation made response and mitigation plans and efforts extremely difficult. 320 km of coastline were heavily oiled up and up to 2100 km somehow felt the effects. This is of course not before it completely poisoned the habitat of salmon sea otter seals and sea birds heavily damaging a fragile ecosystem. Hundreds and thousands of animals were killed including whales and even now nearly 30 years later pockets of crude oil still remain. After the spill, Exxon Valdez as a company changed its name focusing only on oil tanker and ore carrier operations.

John Hazelwood had felony charges pressed against him but was acquitted of them in March 1990 and convicted for a single charge of misdemeanor negligence. He was fined $50,000 and ordered to perform a thousand hours of community service. The months following the oil spill federal responders worked with Exxon employees and over 10,000 Alaskan residents to clean up the oil spill. Exxon Mobil paid 1.8 billion dollars for habitat restoration and 2 billion in cleaning up costs. The beaches were washed with hot water and animals were rescued and cleaned from the oil that trapped them.

Specific areas of the shoreline were left untreated so environmental Studies could be done on the efficiency and effectiveness of the cleanup measures. It was found that high pressure hot water hoses did a good job in removing the oil but intensified the ecological damage by Killing the few remaining plants and animals. One such area is one such area is known as Mearns Rock that is a large shoreline Builder. It was found that the mussels, barnacles, and different kinds of flora had returned to normal levels about four years after the spill occurred. The exact environmental damage of the pristine and fragile wilderness is difficult to estimate but scientists' approximation state 250,000 seabirds were killed together with 3000 otters, 300 seals, 250 bald Eagles, 22 killer whales and countless salmon and herring (Figure 10.8).

Figure 10.8: Sea otters covered in oil from the Exxon Valdez oil disaster.

Source: https://devastatingdisasters.com/exxon-valdez-oil-spill-1989/.

Studies that were done in 2001, 12 years after the oil spill showed that nearly half of the beach sites that were tested in Prince William sound were still contaminated with oil. The 3000 sea otters that died accounted for nearly 50% of the otter population there and they did not recover until

2014, 25 years after. After the oil spill fisherman went bankrupt and many shoreline towns felt the deep economic impact in the following years. The schools of fish that once were a big source of income in the area for many small towns never fully recovered. There are still many ongoing efforts to repair the damage including the Exxon Valdez oil spill trustee Council. The Oil Pollution Act of 1990 was passed by President George Bush and was probably the biggest new regulation among others made to ensure a disaster of such a kind Never Happens Again. The act first of all and shows than any oil tankers in United States Waters must have a double how and it also increases the penalties for corporations that are responsible for oil spills. The Exxon Valdez ship was sold repaired and returned to service under a different name and can no longer travel in US Waters because of the new regulations. The European Union also adopted similar stringent regulations but a little later on in 2002.

10.8. THE LIBBY, MONTANA ASBESTOS TRAGEDY (1999)

The town of Libby Montana has an economic history largely supported by its natural resources through logging and mining. In the early in the early 1900s and mineral known as vermiculite was discovered in the mountainous region and a chemical conglomerate known as W.R. Grace Company monopolized the local mines and began producing about 80% of the world vermiculite. Vermiculite is a type of naturally occurring mineral that when heated expand and has a low density and thermal conductivity. It is used in gardening and agriculture, packing material in the automotive industry for friction lining, in construction for the manufacturing of building boards, plasters, and loose fills insulation. Not all vermiculite contains asbestos but that of the local mines did. Nowadays Minds across the world are regularly tested for it but old products made with vermiculite do contain some amounts of the silicate. Asbestos is a naturally occurring silicate mineral that comes in long and thin fibrous crystals when observed under a microscope.

This ability for asbestos to break down microscopically into thin fibers makes it very dangerous and harmful to one's health because once lodged in lung tissue it causes cancer asbestosis and mesothelioma. W.R. Grace and Company was aware of some of the dangers and even discussed the risks of vermiculite mining in his private memos but to sustain profits he did not inform his miners or the town's residents and discarded wastes on to playgrounds, fields, and roads. The ecological effects of asbestos come down to its size because it cannot be detected with the naked eye and contaminates the air water and soil causing health hazards to both humans and animals alike.

By 2010 more than 200 people had died which constitutes 10% of Libby's population and 2000 severe illnesses were reported. The situation the situation was so bad that the Environmental Protection Agency actually replaced WR Grace Company as the number one employer in the town. They were charged with 250 million dollars to clean up 1890 residents that were contaminated and were subjected to 270,000 lawsuits of which 120,000 have not been paid yet. The town of Libby is the first public health emergency in US history and any residents in the town diagnosed with asbestos-related diseases are eligible for full single-payer Medicare coverage provided by the government of the United States (Figure 10.9).

Figure 10.9: Vermiculite from the mines of Libby, Montana containing asbestos.

Source: https://www.asbestos.com/news/2018/01/04/libby-amphibole-asbestos-disease/.

10.9. HURRICANE KATRINA (2005)

Hurricane Katrina was the most publicized hurricane of the century and in relation to economic costs the only comparable to Hurricane Harvey of

2017 costing 125 billion dollars in damage. Its true effects were felt most intensely in the city of New Orleans and the adjacent areas. According to the National Oceanic Atmospheric Administration (NOAA), it was the fourth most extreme Atlantic hurricane on record to make landfall on continental United States. The three hurricanes that beat it in terms of strength were the 1935 Labor Day hurricane 1969 hurricane Camille and the 2018 hurricane Michael. Hurricane season of 2005 was in fact the most active hurricane season ever recorded with 31 tropical and subtropical cyclones on record. Tropical storms are usually given names in the order they come in accordance to the alphabet and that year meant that the list was exhausted and hurricane Katrina was the worst of them (Figure 10.10).

Figure 10.10: High resolution photograph by NOAA and NASA of Hurricane Katrina shortly after landfall.

Source: https://www.flickr.com/photos/gsfc/20471874261/in/album-72157656646633089/.

"What initially started as a natural disaster quickly became a natural disaster and a failure of government to look out for its citizens" these are the words of President Barack Obama as recorded by the BBC. The reason that hurricane Katrina counts as a human-made disaster and not a natural one is because of the subsequent flooding of New Orleans after the hurricane itself. The levees and the Flood protection system that the city had built were poorly designed and maintained and to make matters worse the waterway systems that were engineered up and down the Mississippi River to facilitate navigation to and from the Gulf of Mexico where also major failures in the part of infrastructure policies and engineering decisions made by those in charge.

At the time of the hurricane in New Orleans was lying an average of 6 feet below sea levels with many neighborhoods as low as 11 feet below sea level and the primary line of Defense against the sea and the Mississippi River was a massive levy and flood wall system. For this city to have been hosting half a million homes with the levee and flood protection system that's received little to no Investments over half a century prior to the hurricane was a recipe for disaster. The failure for proper development and infrastructure policies together with the aforementioned storyline is the reason why hurricane Katrina turned from a natural disaster into a man-made catastrophe. The tropical depression first developed on the 23rd of August across the Atlantic and became a tropical storm the following day and a hurricane after that on the 25th. It crossed over the Gulf of Mexico quickly intensifying into a category 5 hurricanes as it made landfall in Southern Florida on the 25th. After making landfall, the storm weakened as it veered north but still had winds of up to 200 km per hour that hit Southeastern Louisiana on the 29th. The levees and flood protection system had 50 separate failures which ended up flooding most of New Orleans transportation and communication facilities that left tens of thousands of people stranded without food shelter basic necessities.

1200 people lost their lives. The scale of the disaster was so massive that it provoked national and international response efforts and on top of that federal local and private rescue operations. Multiple investigations were held After the Storm and it was concluded that the US Army Corps of engineers was responsible for the levee and flood control systems failures. We were designated and built the region's levies almost five decades earlier and were responsible for maintaining it but did not. In actual sense, they were liable for all the financial damage but because of the flood control act of 1928 the sovereign immunity clause that dictates a sovereign or state cannot commit legal wrong and his immune from civil suits and criminal prosecution allowed them to be unscrutinized.

The subsequent storm ripped across the south and eastern coast of the United States having damage as far as Kentucky and even up to Canada. In fact, a 10-year-old girl was reported to have drowned in the heavy rainfall and flooding of Hopkinsville, Kentucky. Severe local flooding occurred even as far north as Quebec prompting evacuations as rivers broke their Banks and sewage systems well overwhelmed. Over 1 million people were displaced many moving to Alabama Texas and Illinois and the state of Louisiana display the population decline of nearly 5% (almost 220,000 people)

From an environmental perspective, Katrina was nothing less than a catastrophe with substantial and extensive beach erosion. Several barrier Islands like Dauphin Island and the Chandeleur Island were actually pushed towards land and were severely damaged. Over 550 square kilometers of land what taken by The Waters of both hurricane Katrina and hurricane Rita that followed. The wetland areas and the bayous of Louisiana were already suffering from subsidence which is when the ground surface sinks and plenty of erosion and hurricane Katrina only made the situation more severe. Is area supported the lives of many marine mammal pelicans Turtles and fish and some migratory species like the redhead ducks. Nearly 20 different National Wildlife refuges and sanctuaries had to be shut down entirely from the damage

On top of that, Katrina ripped apart a large amount of trees along the Gulf Coast. The standard mortality rate for the trees in the area before the storm was nearly 2% but increase to 20% by the end of the year. The ecosystem was so badly damaged and by fact, there was an increase in carbon emissions. To make the situation even more distressing the storm caused multiple oil spills from nearly 50 different facilities with the total of over 7 million gallons oil spilt. The residual flood waters not only mixed with the massive amount of leaked oil but also contained a large amount of raw sewage bacteria pesticides and toxic chemicals together with dangerous heavy metals.

To get an idea of the mental and psychological effects of hurricane Katrina there are two scientific studies we can take to account. The first was a study was done by the community mental health journal on the hurricane Katrina evacuees in Texas and almost half of them fit the criteria of suffering depressive disorders and PTSD. Another study published by the maternal and child health journal show that women that were pregnant and exposed to the terrors of hurricane Katrina had given birth to children that were underweight on average. The study concluded that women who are vulnerable to disasters become more valuable to poor pregnancy terms.

10.10. KINGSTON FOSSIL PLANT COAL FLY ASH SPILL (2008)

The Kingston coal spill accident was an environmental and industrial catastrophe in 2008 that occurred because a dam that was holding 1.1 billion gallons of coal ash ruptured in Tennessee Valley. The spill covered an area of 1.2 km squared in the surrounding land damaging several homes and flowing into the Emory River and Clinch River that are both tributaries of the larger Tennessee River. It is ranked as the worst call Ash disaster in US history and the largest fly Ash release. The plant would generate fly Ash from the Burning of finely ground coal to produce power and dispose of the residues in an ash pond just north of the plants along the Emory river.

Fly Ash consists of fine particulate matter which rather than releasing it into the atmosphere is ordinarily mixed with water and pumped into a retaining residue pond where the overtime will gradually settle to the bottom and then reached out and moved to drying cells. This particular point had some dykes and dams that would impound this watery slurry. The breach in the dam occurred at the North West End of the ash pond on Monday December 22 2008 releasing 4.2 million cubic meters of toxic industrial waste.

The cause of the disasters was investigated and reported to have been as the result of a spillage of an unstable layer of wet coal ash directly underneath the pond. The terrace lining was also positioned in a way that it narrowed the area for storing and increased the pressure exerted on the specific area of the dam buy rising stacks. Between December 1st and December 22nd when the disaster happened, the area received a total of 16.4 centimeters of rain and that combined with below-freezing temperatures contributed to the failure of the embankment. The spill occurred over the course of about an hour in consecutive waves covering the surrounding land with up to six feet of sludge and slurry.

Thankfully, nobody died in the immediate spill as the surrounding areas of the power plant was largely rural and not residential. However, it is reported that several employees eventually developed illnesses like brain cancer and leukemia as a result the exposure of the toxic industrial waste and 10 years after the disaster it was reported that 40 people had died. Coal may not be classified as a hazardous material but contains heavy metals that cause cancer liver damage many neurological diseases and other health issues.

The mud flow wave completely covered 12 homes and ripped one off entirely from its Foundation with 22 residents having to be evacuated though none was injured. 42 residential properties were damaged, a major gas line was ruptured, and a water main was broken as well as several power lines. The slurry washed out of roads and obstructed the rail line that was meant for trains to deliver coal to the plant. To put it into perspective the amount of toxic industrial waste that was spilled was enough to fill 1660 Olympic-sized swimming pools and was over 100 times larger than the 1989 Exxon Valdez oil spill already discussed above (Figure 10.11).

Figure 10.11: The fly ash slurry entering the Emory River.

Source: https://insideclimatenews.org/news/04122018/toxic-coal-ash-spill-illness-verdict-kingston-tennessee-cleanup-workers-compensation.

The environmental damage was without question as countless trees were downed and innumerable fish were killed. The Environmental Protection Agency together with the Tennessee department for environment conservation tested the water quality. The river waters displayed elevated levels of Lead thallium Mercury and arsenic together with many toxic chemicals like copper barium cadmium nickel and Chromium. The spill ignited debate about the regulation of the coal industry and coal ash treatment across the US. Lisa Jackson the head of the Environmental Protection Agency under Barack Obama's administration immediately called for review on the coal ash disposal sites in the country and a bill was introduced to regulate coal ash disposal.

The cleanup efforts were estimated to take up to six weeks however after 6 months only 3% of the spill was cleaned and it costs 975 million dollars. The cleanup was in accelerated with efforts from the EPA and the TVA where the actual safely transported to a permanent line leachate collecting facility in Alabama called arrowhead landfill and the final phase of the cleanup was completed in 2015 and the total cost was approximately 1.134 billion. By 2017, it was reported that the affected ecosystem had returned to its conditions prior to the spill. A new dam and protective levee was built with an earthquake-proof clay layer and the company bought over 960 acres and built a park on the area where several homes were damaged by the spill. This disaster brought a lot of reflective thought and legislation around the environmental impact of the coal industry

10.11. THE DEEPWATER HORIZON (BP) OIL SPILL (2010)

The Gulf of Mexico has suffered very many environmental and man-made disasters but the Deepwater Horizon oil spill is one for the record books. The Mississippi Canyon Block 252 (Mocando Prospect) was an oil and gas hydrocarbon exploration project by the United States in its exclusive economic zone off the coast of Louisiana. The British Petroleum company Plc. (BP), a multinational oil and gas company based in London was the main operator and primary developer of the oilfield with 90% interest. Surveys were first conducted in 1998 and BP purchased the mining rights to drill in the area in March 2008.

The first drilling began in October 2009 with a massive Transocean Marianas drilling unit that is semi-submersible and can go up to depths of 7000 feet and is paired with a 15000 pounds per square inch blowout preventer that it is a specialized valve used to control and monitor oil and gas as it comes from below the surface of the Earth. It was constructed by a Hyundai Heavy Industries and was named Deepwater Horizon. The rake was 10 years old but in great shape when it was, set up however drilling was suspended between November 2009 and February 2010 due to damage to the rig caused by hurricane Ida.

On April 10th at approximately 7:45 p.m. methane gas extracted under high pressure expanded into the drilling riser where it ignited and exploded at the top of the drilling rig engulfing the platform. There are 126 crew members on board and 11 lost their lives and we never actually found even despite a continuous three-day search by the US Coast Guard. Immediately rescue operations began first to assist the 17 that were injured and evacuate the rest as well as put out the inferno that continued to burn. The drilling riser was injected with drilling mud to counteract the pressure of oil and natural gas escaping but it continued to discharge oil into the Gulf. The oil leak was actually discovered on the afternoon of April 22nd 2 days after the explosion and when a large oil slick was observed spreading around the former rig site.

Oil continued to flow for an unbelievable 87 days after that. BP estimated the rate flow of oil to be about 5000 barrels per day at the most but the US government released reports that this may have picked her up to 60,000 barrels per day. The total estimated value of oil that looked across the Gulf of Mexico was a mind-blowing 4.9 million barrels with a 10% margin of error. was buy good deal the world's largest accidental oil spill and BP obviously challenged the U.S. government stating they overestimated the volume saying that they did not account for the word that was collected or that which was burnt before it could enter the gulf waters. The spill covered an area of 180,000 square kilometers of Ocean according to satellite images which is comparable to the size of the state of Oklahoma. Tar balls together with oil Sheen and trails who recorded up until January the following year in various parts of Louisiana and Mississippi at large. In 2013, 3 years later the effects were still being felt as oil covered materials were still being recovered from the Louisiana coast.

Immediately the disaster happened there were attempts to stem the flow of oil. The first was BP trying to shut the blowout preventer which is the specialized industrial valve but this is done on successfully. Next day placed a containment Dome that was meant to cover the largest leak and transfer the oil into a storage vessel. On shallow water and slicks, this worked however; there were methane hydrate crystals that blocked the openings within the pipe. The next attempt was a top kill witches a procedure where heavy drilling fluids are pumped into the leaking pipes to stop the flow of oil before sealing. All these attempts failed. After this, a riser in session 2 was employed and it collected almost 1 million U.S. gallons of gas onto a drill ship known as discoverer Enterprise. This at least was 1 million gallons that was not floating around in the sea. There were recurrent or continued leakages recorded a couple of years later but they were comparable in size to naturally occurring oil seeps so they did not endanger or threaten the wildlife. An oil dispersant known as coexist was also used and it contains a mixture of solvents and emulsifiers that separates the oil into tiny droplets and this method was actually termed purely experimental (Figure 10.12).

Figure 10.12: The deepwater horizon rig on fire.

Source: https://www.britannica.com/event/Deepwater-Horizon-oil-spill#ref294125.

The economic costs where in part felt by BP but lost tourism and fishing around the Gulf Coast meant that up to 22 billion dollars was lost up until 2013. The state of Louisiana lost 153 million dollars alone from the lost tourist activity and regions commercial fishing industry lost 2.5 billion dollars

The area where the Deepwater Horizon oil spill occurred showed damage to 1000 different species of animals and the environmental impact was beyond belief. They polycyclic aromatic hydrocarbons (PAHs) that are carcinogens and cause several health risks not only to human beings but animals and in this case marine life as well. The vicinity of the oil spill was actually termed as one of the most bio diverse ecosystems on the planet and PAH, levels were studied by the national oceanic and atmospheric administration together with Stanford University and the Monterey Bay Aquarium. Fish and the study specifically focused on Bluefin tuna that were exposed to the spill developed all sorts of deformities of the heart and general internal organs that either shorten their lifespan or prove to be fatal all together. Oil that sunk to the Ocean Floor and endangered species of

Louisiana pancake batfish that are indigenous only to that area. Traces of oil and the Corexit dispersant that were used were found in blue Crab lover and essentially, it entered every level of the food chain. The biodegraded area also meant that the oxygen levels in the water was drastically reduced and even pelican eggs were found to contain petroleum related compounds. There was a large number of mutated fishes with several species of shrimp found locking eyes and eye sockets. Dead baby dolphins kept washing up along the coast of Mississippi and Alabama in disturbing numbers almost 10 times more than natural. Some dolphins were even born stillborn or with very abnormal or underdeveloped lungs. The ecosystem around the Gulf Coast is still poisoned and only just now gradually beginning to recovering over a decade later and continuing studies still show distressing results (Figure 10.13).

Figure 10.13: Oil from the deep water horizon accident washing up on the coast of Louisiana.

Source: https://en.wikipedia.org/wiki/Deepwater_Horizon_oil_spill#Deepwater_Horizon_drilling_rig.

Bibliography

1. Anthony, E., & Eric, L. H., (2016). The Consequences of Global Disasters (p. 252). London: Routledge.
2. Babar, M., (2007). Environmental Changes and Natural Disasters (pp. 1–73). New Delhi: New India Publishing.
3. Battersby, S., (2013). Clay's Handbook of Environmental Health (pp. 1–85). London: Routledge.
4. Birkmann, (2007). Measuring Vulnerability to Natural Hazards: Towards Disaster Resilient Societies (p. 524). New Delhi, The Energy and Resources Institute (TERI).
5. Coppola, D. P., (2010). Introduction to International Disaster Management (pp. 1–37). Netherlands: Elsevier.
6. Cutter, S. L., (2012). Hazards Vulnerability and Environmental Justice (pp. 69–151). London: Routledge.
7. Dufresne, E., (2019). Environmental Disasters (pp. 6–20). New York: Greenhaven Publishing LLC.
8. Emerald Group Publishing Limited, (2015). New Perspectives in Global Environmental Disasters (pp. 1–95). London: Emerald Group Publishing.
9. Gallant, B., (2008). The Facility Manager's Guide to Environmental Health and Safety (pp. 1–97). Nairobi, Kenya: Government Institutes.
10. Geipel, R., (2012). Long-Term Consequences of Disasters: The Reconstruction of Friuli, Italy, in its International Context, 1976–1988 (p. 171). Berlin: Springer Science & Business Media.
11. Helen, J., & Douglas, P., (2015). The Consequences of Disasters: Demographic, Planning, and Policy Implications (pp. 3–65). New

York: Charles C Thomas Publisher.

12. IFRC, (2011). Disasters in Africa: The Case for Legal Preparedness. Geneva: International Federation of Red Cross and Red Crescent Societies.
13. Kapur, A., (2010). Vulnerable India: A Geographical Study of Disasters (pp. 3–28). New Delhi: SAGE Publications India.
14. Masie, D., (2020). Insuring Africa Against Natural Disasters. Retrieved from: African Business Magazine: https://africanbusinessmagazine.com/sectors/finance/insuring-africa-against-natural-disasters/ (accessed on 6 June 2020).
15. Meeme, V., (2020). Digital Tools Help Africa Mitigate Climate Disasters. Retrieved from: scidev.net: https://www.scidev.net/sub-saharan-africa/climate-change/news/digital-tools-help-africa-mitigate-climate-disasters.html?__cf_chl_jschl_tk__=acee45d7bb1ec20f8d87e7e4a423d165a50b808b-1587369155-0-Aa-SQ98PxXqhhFpngUEQSVgYyqGLlG4YLaT9IlDiBkVbD40xKgvbaZMXSteyx (accessed on 6 June 2020).
16. OCHA Services, (2019). 2019: Natural Disasters Claim More Than 1200 Lives Across East and Southern Africa. Retrieved from: reliefweb.
17. Paul, B. K., (2011). Environmental Hazards and Disasters: Contexts, Perspectives, and Management (p. 352). Florida: John Wiley & Sons.
18. Pelling, M., (2012). The Vulnerability of Cities: Natural Disasters and Social Resilience (p. 224). London: Routledge.
19. Ricci, P., (2006). Environmental and Health Risk Assessment and Management: Principles and Practices (pp. 39–70). Berlin: Springer Science & Business Media.
20. Simelane & Thokozani, (2019). Natural and Human-Induced Hazards and Disasters in Africa (pp. 1–15). Cape Town: Africa Institute of South Africa.
21. Smith, K., (2003). Environmental Hazards: Assessing Risk and Reducing Disaster (pp. 1–79). London: Routledge.
22. Srinivas, H., (2015). Environmental management and disaster reduction: An introduction. Concept Note Series E-021, 1–8.
23. Taylor, B., (2005). Effective Environmental, Health, and Safety Management Using the Team Approach (pp. 1–89). Florida: John Wiley & Sons.

24. United Nations, (2010). Natural Hazards, Unnatural Disasters: The Economics of Effective Prevention (p. 254). London: World Bank Publications.
25. Veenema, T. G., (2003). Disaster Nursing and Emergency Preparedness for Chemical, Biological, and Radiological Terrorism and Other Hazards (p. 586). New York: Springer Publishing Company.
26. Weil, A., (2004). Environmental Disasters (pp. 6–12). New York: Saddleback Educational Publ.
27. World Bank Group, (2017). Post-Disaster Recovery Plan (pp. 1–162). Sri Lanka, Floods and Landslides.
28. Worthington, D., (2003). Dictionary of Environmental Health (p. 304). London: Routledge.

INDEX

C

D

F

G

H

I

Printed in the United States
By Bookmasters